# 285 Broken Dreams

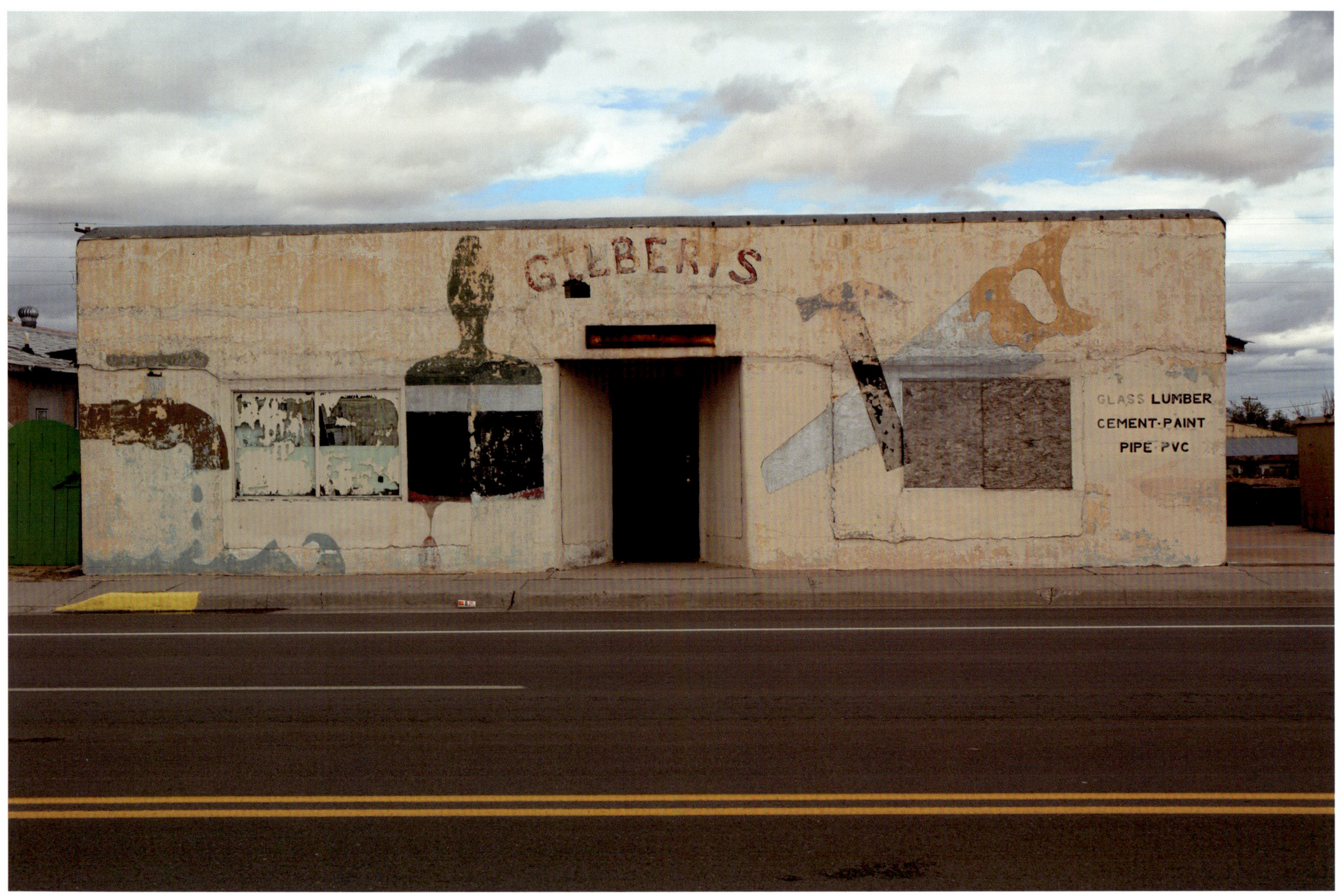

GILBERTS
GLASS LUMBER
CEMENT·PAINT
PIPE·PVC

# 285 Broken Dreams

PHOTOGRAPHING SOUTHEAST NEW MEXICO TO TEXAS

BY CHRIS ENOS

WITH AN ESSAY BY ELVIS E. FLEMING

MUSEUM OF NEW MEXICO PRESS
SANTA FE

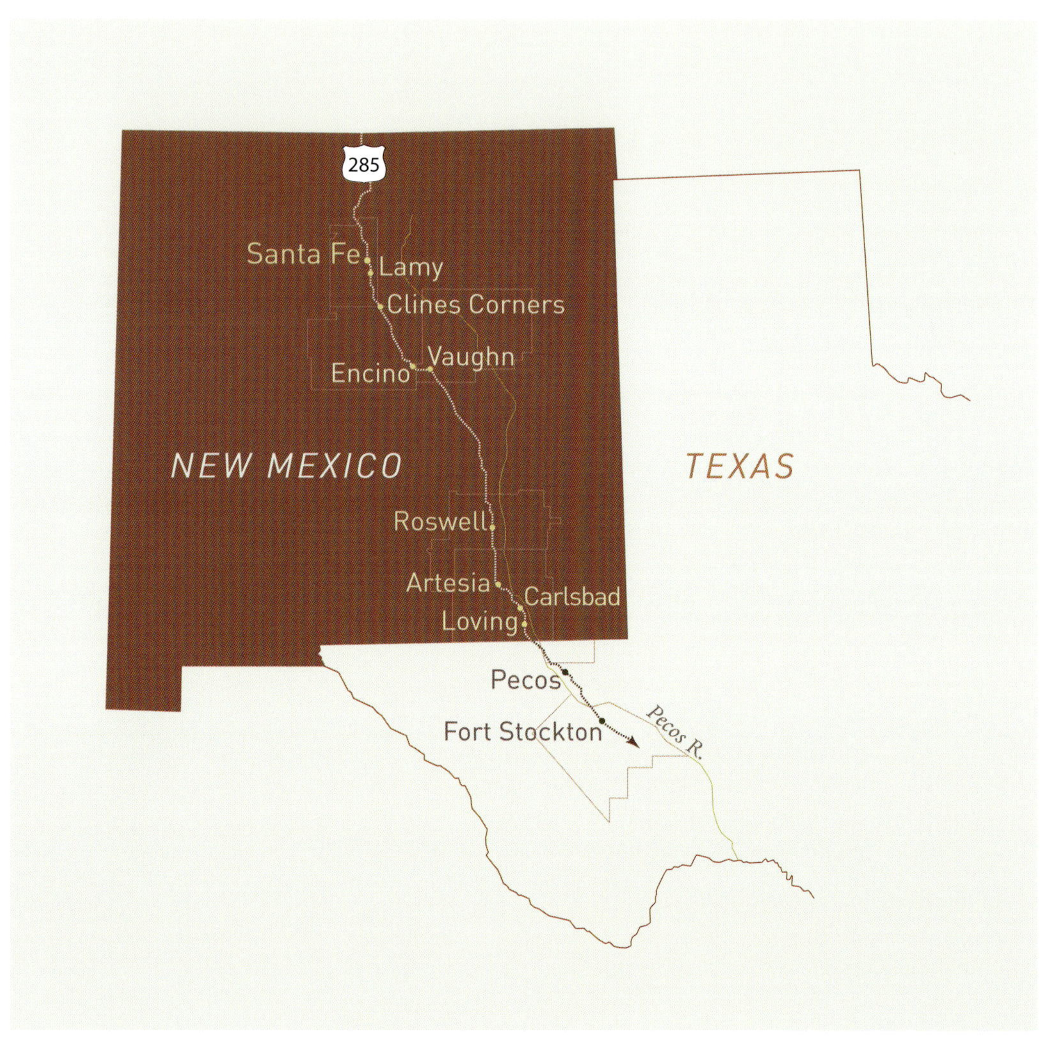

285
Santa Fe
Lamy
Clines Corners
Vaughn
Encino
NEW MEXICO
TEXAS
Roswell
Artesia
Carlsbad
Loving
Pecos
Fort Stockton
Pecos R.

## Preface



MY FIRST TRIP SOUTH on U.S. Route 285 was in the summer of 2009. I had planned a camping trip to Big Bend National Park in Texas for the spring but was delayed by illness. In early summer I packed up my sixteen-foot Casita trailer and two dogs and headed south from Santa Fe.

Road trips have always been a way for me to take a break from the self-centered daily stuff of life and concentrate on the world around me. These are the times that I can just look and observe and think. After a while my focus naturally goes to something that is relevant to my life at the time.

I didn't have the creative energy to photograph on this trip, but I began to concentrate on the decaying homes and businesses along the highway that I was attracted to. Very aware of the multitude of photographs of decaying buildings and ghost towns in many places, my first thoughts were that this is an over-photographed subject and I did not need to contribute more. But what I was seeing seemed particular to this highway and corridor on the east side of New Mexico and into west Texas. There was a character to the buildings, businesses, and homes. People had clearly made individual choices as to how these buildings were constructed and decorated—to build in wood, plaster, or metal, what shape of roof, the use of color. The architecture was spare and suggested a tough life with no frills. Buildings built by rugged people with little means. No swimming pools and well-kempt lawns here, just the essentials.

My second trip the following fall I was determined to find out what it was that compelled me to take this mostly very boring and dreary trip again. When I got to Encino and started to photograph the old buildings, I began to think about the people who moved to this rural area to begin a new life. They had to be tough to live in this desolation. I thought about the people arriving, building a home and/or business and then having to leave. I thought mostly of their broken dreams.

This became the focus of subsequent trips along route 285 from my home in Santa Fe to Fort Stockton, Texas. It became a metaphor for my life at the time.

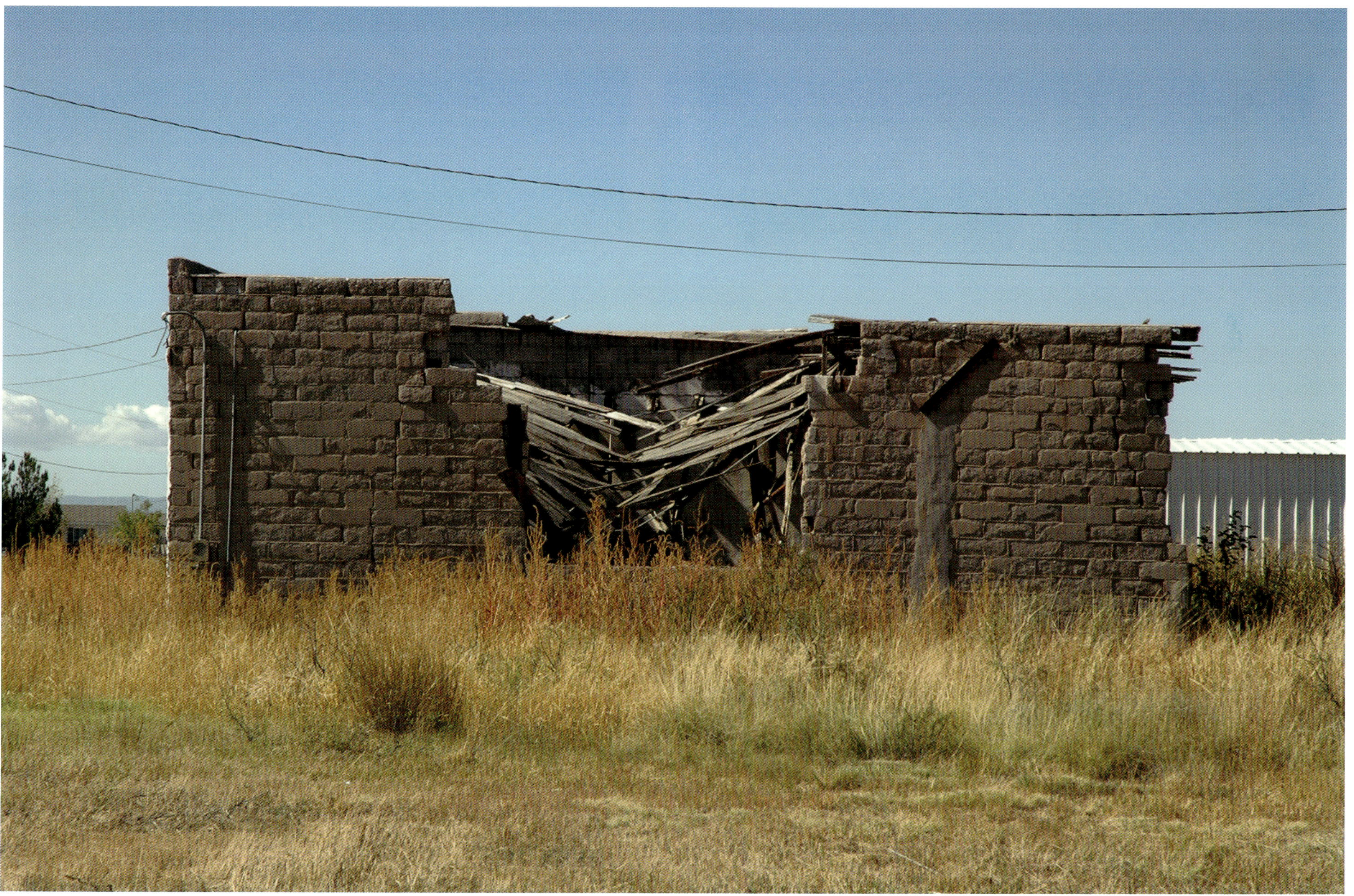

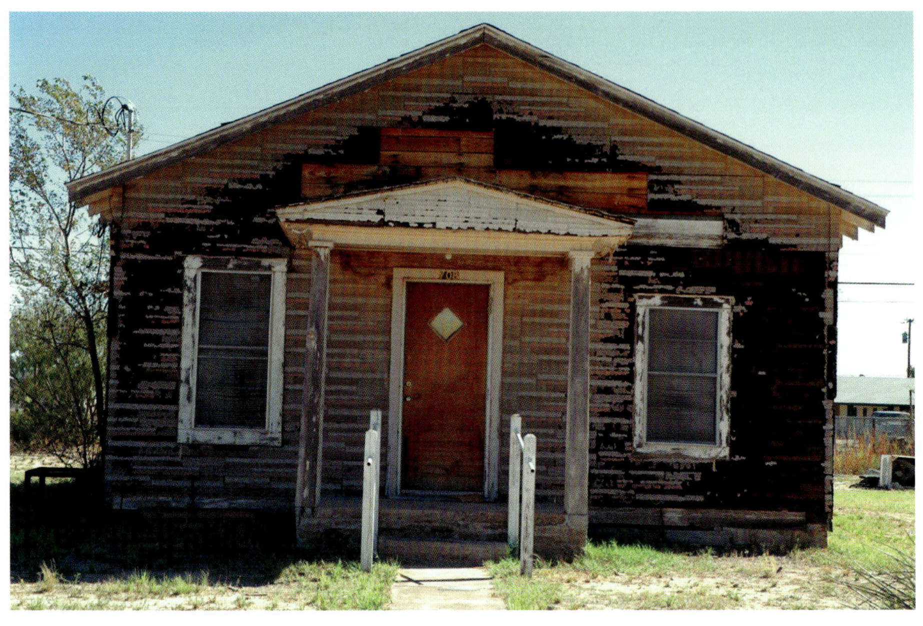 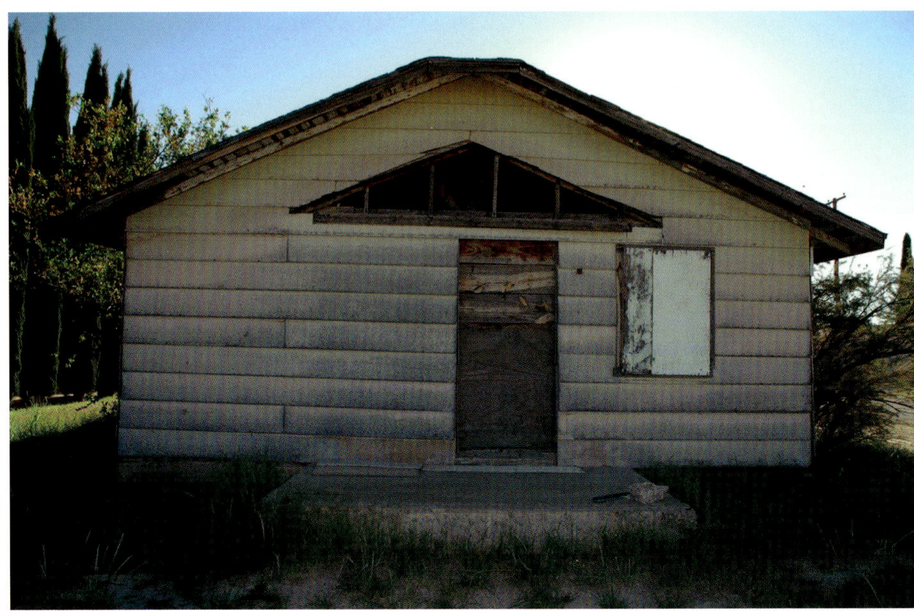

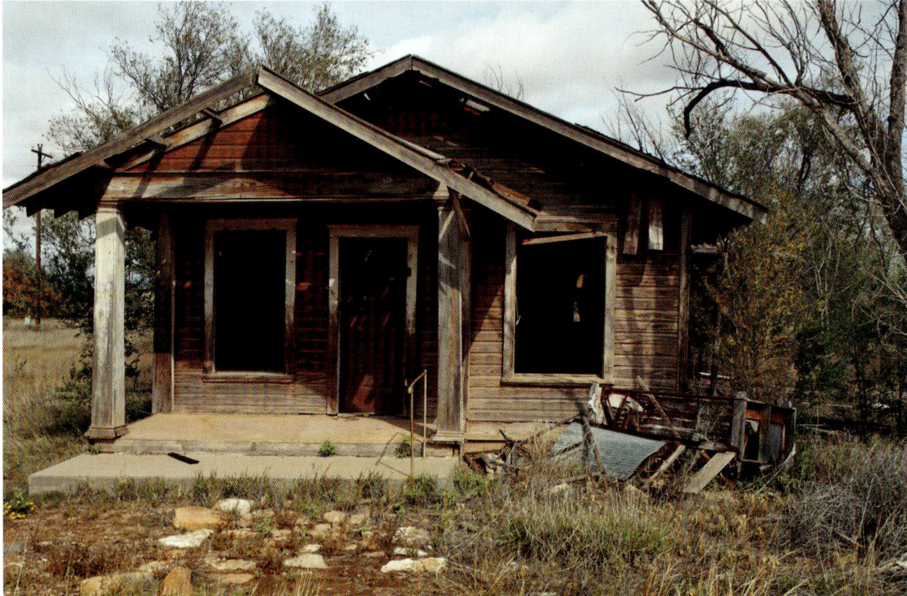 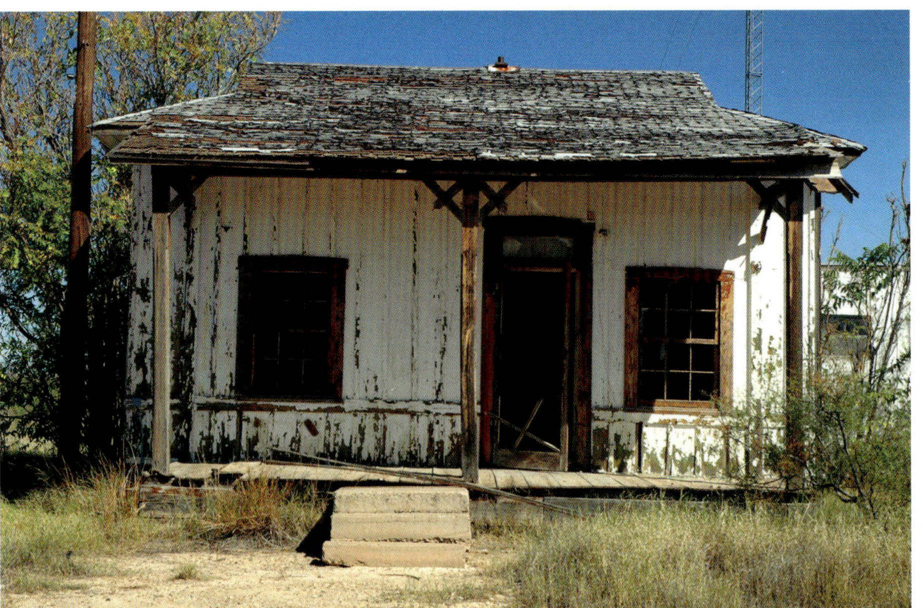

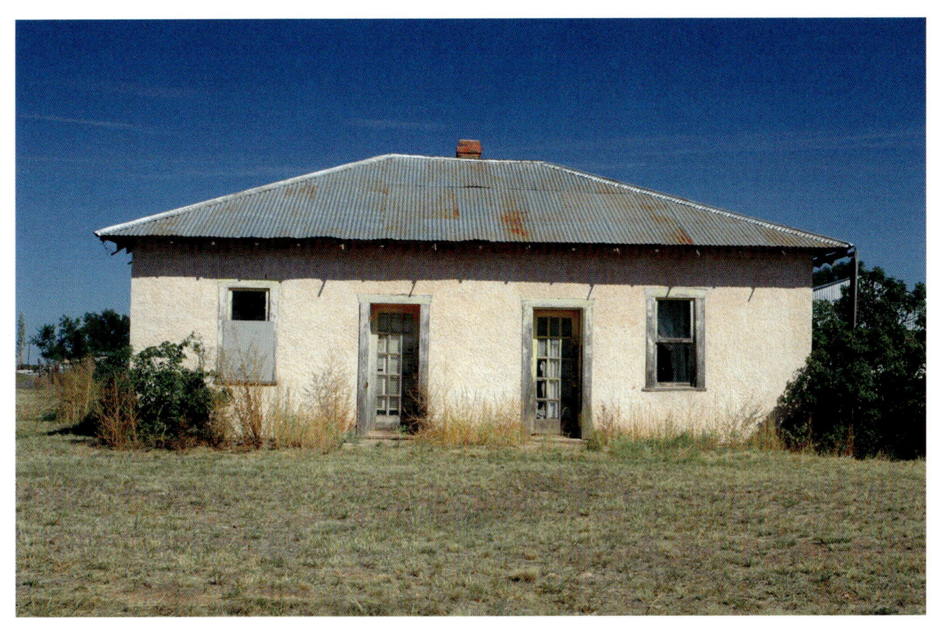
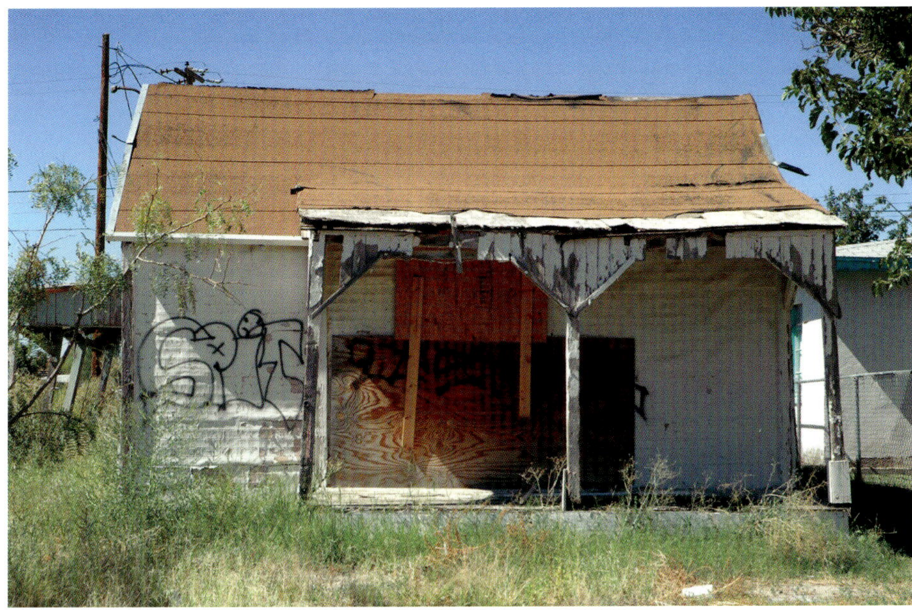
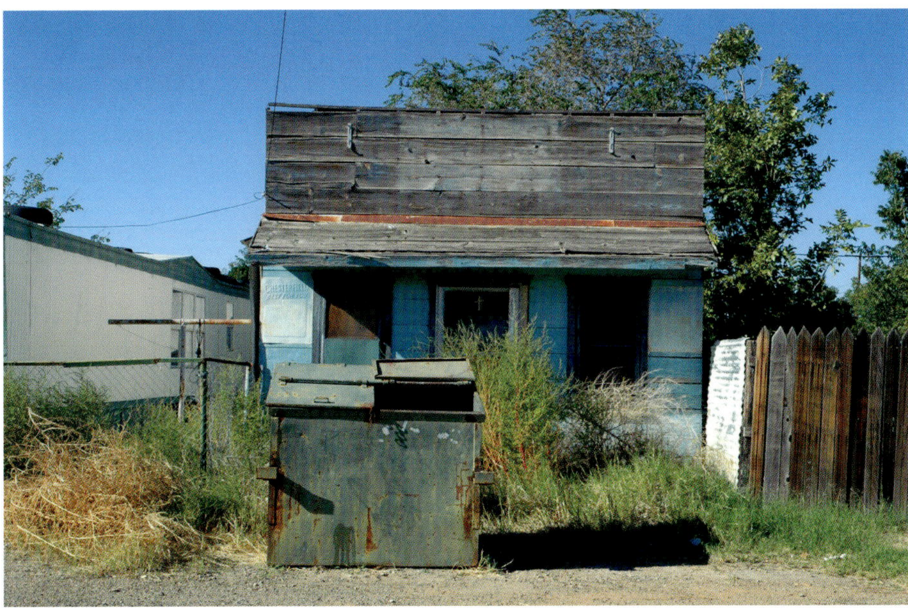
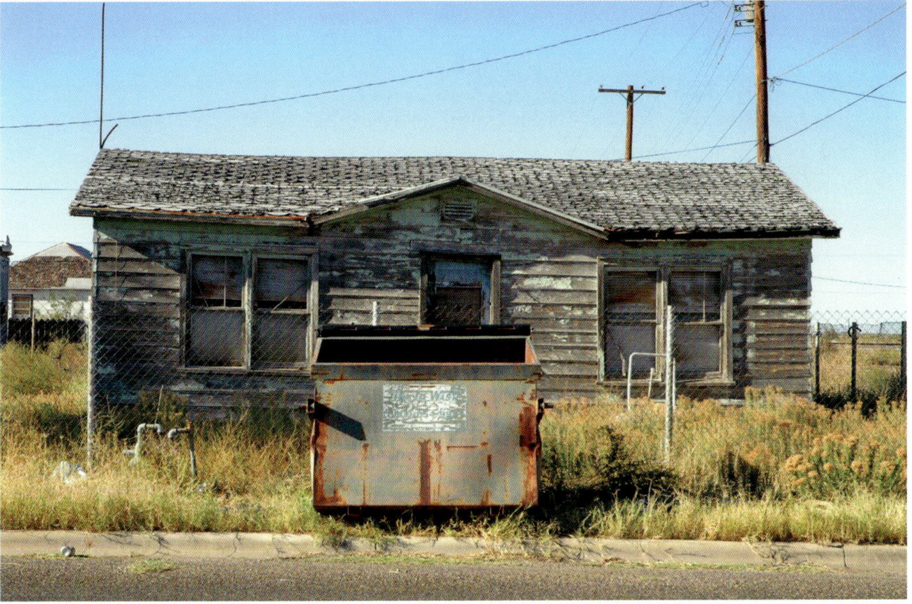

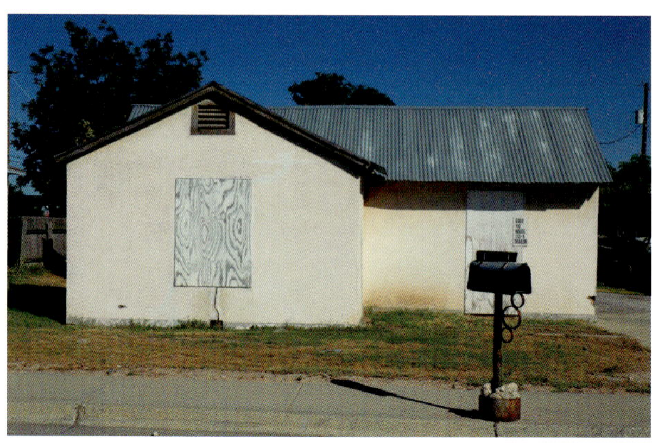 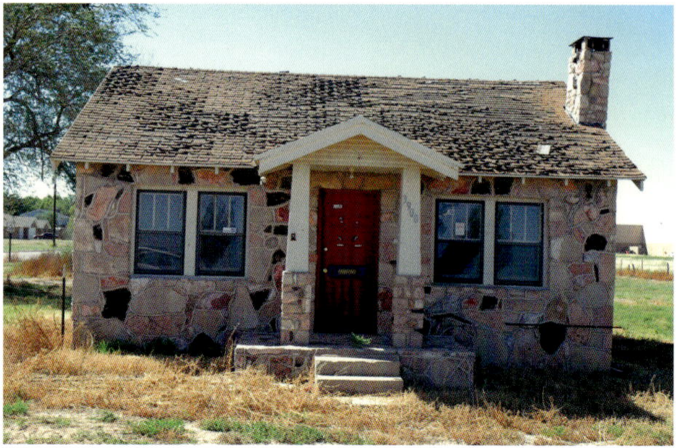 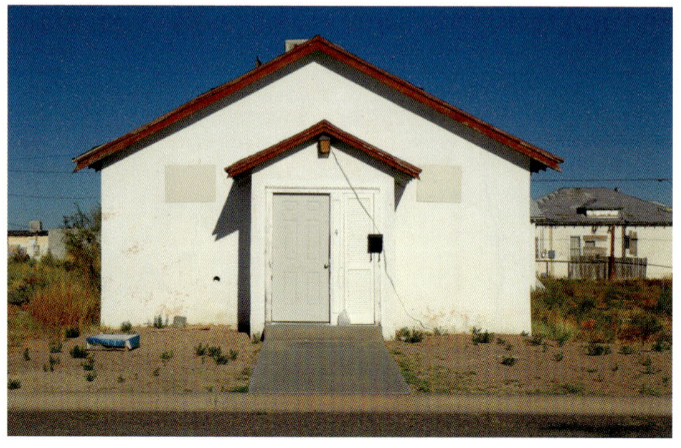

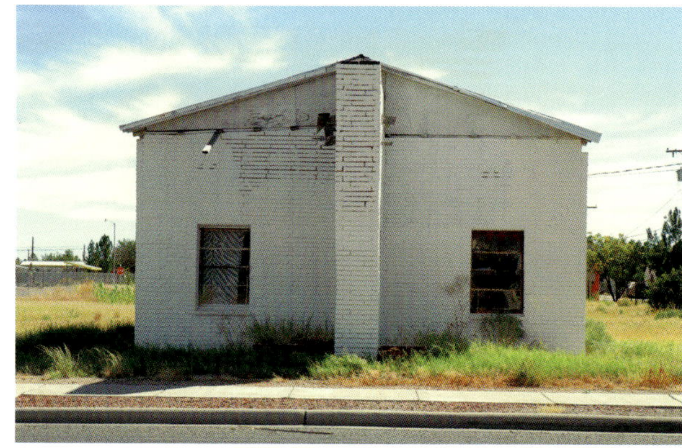 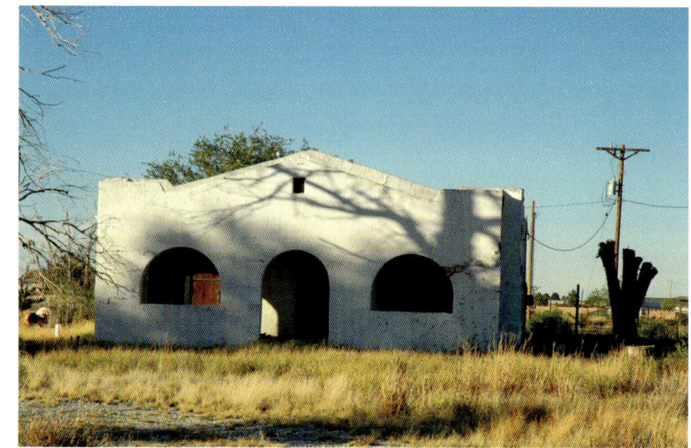 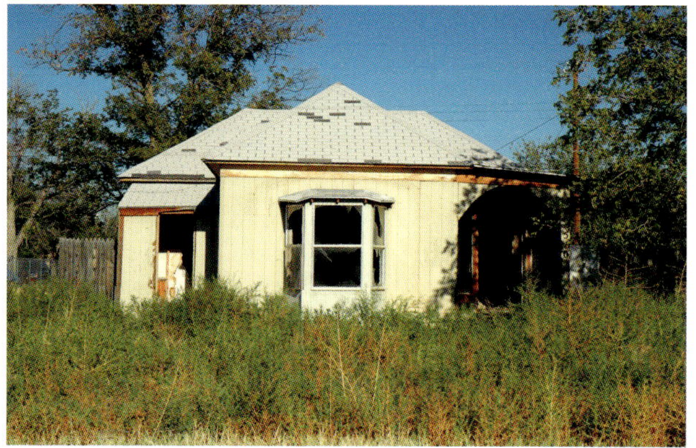

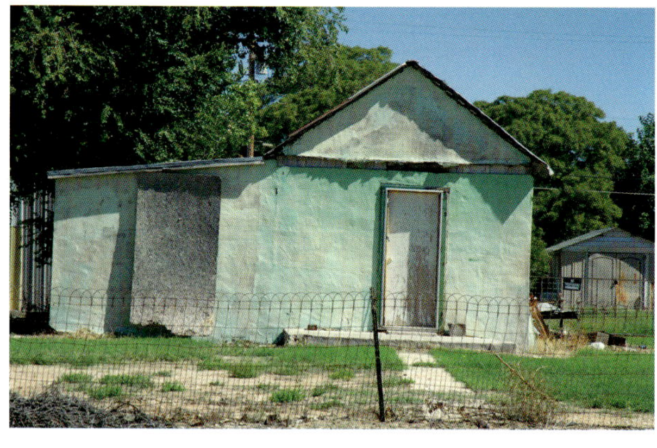 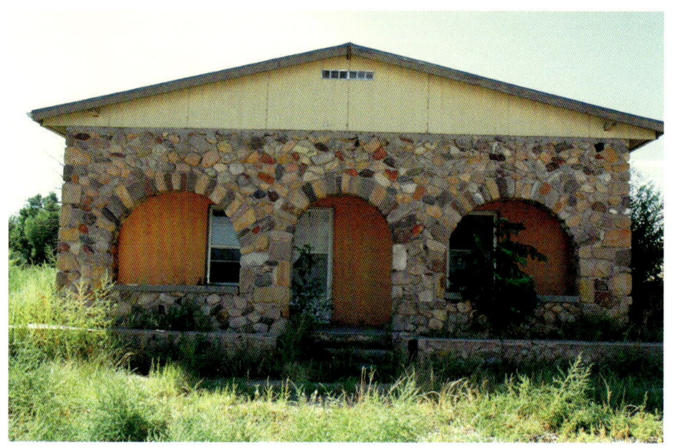 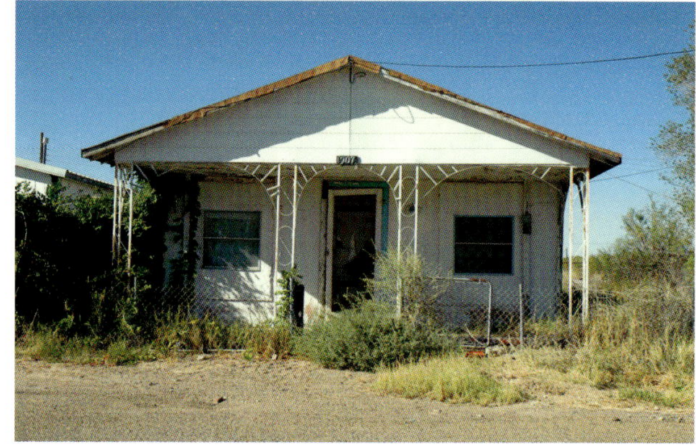

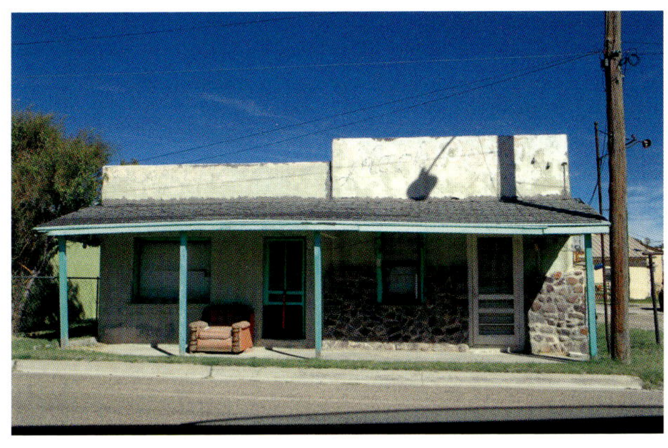
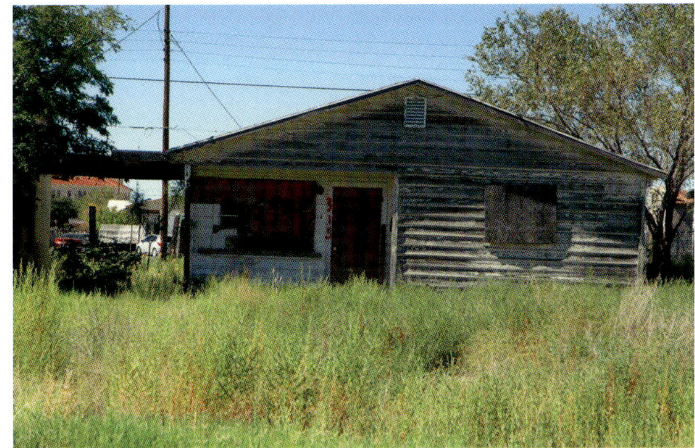
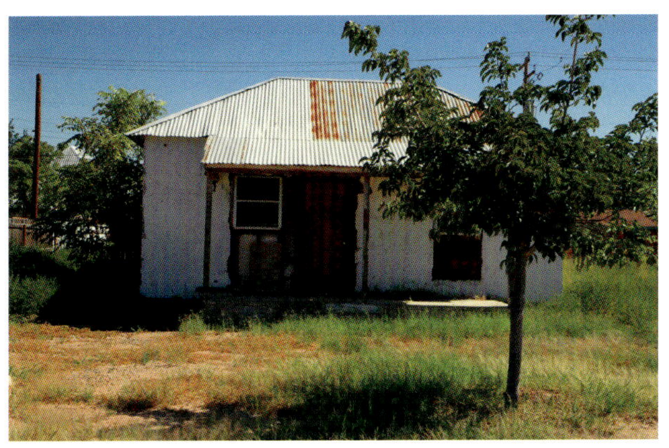
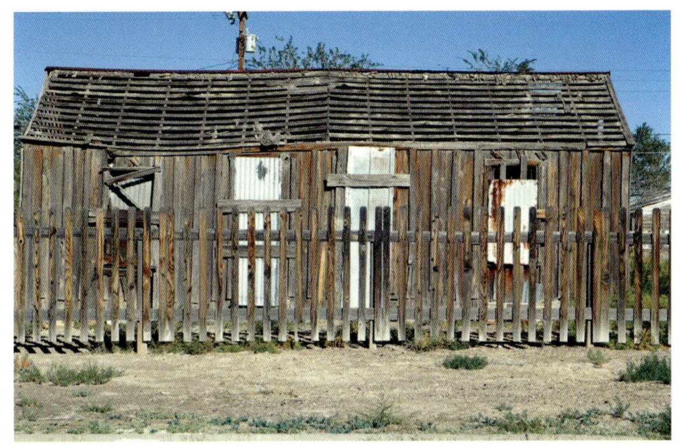
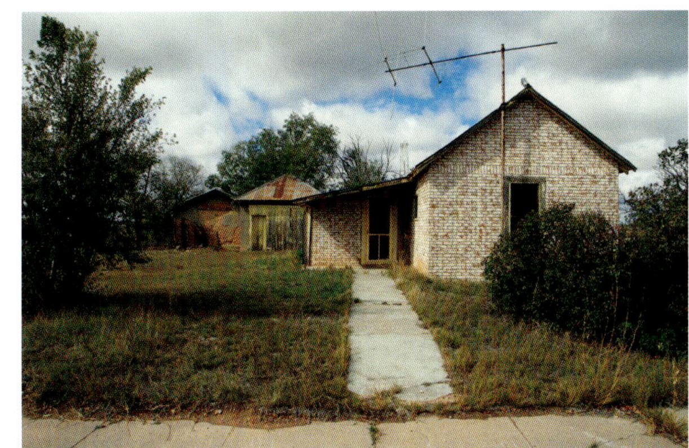
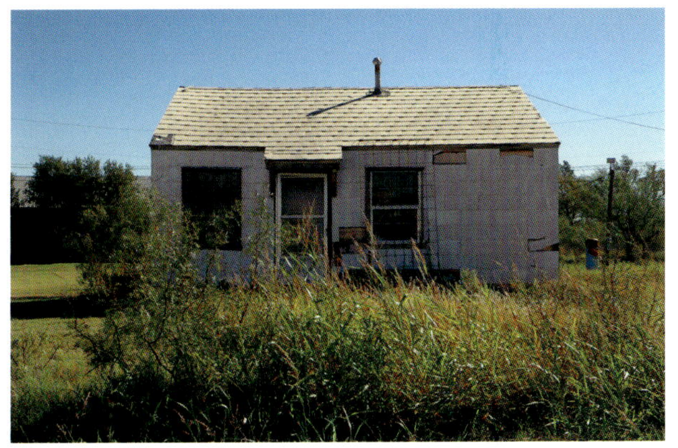
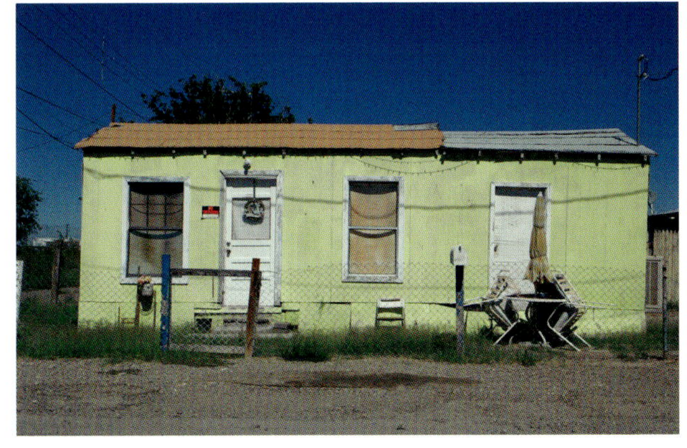
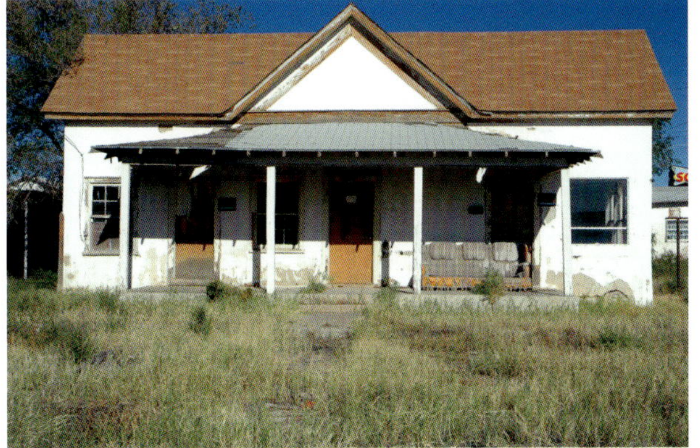
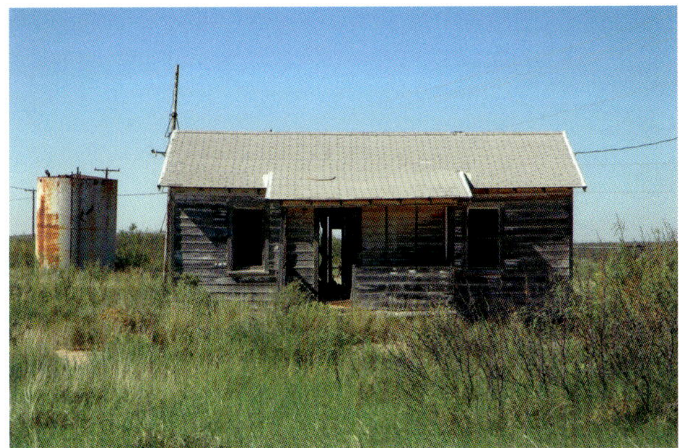

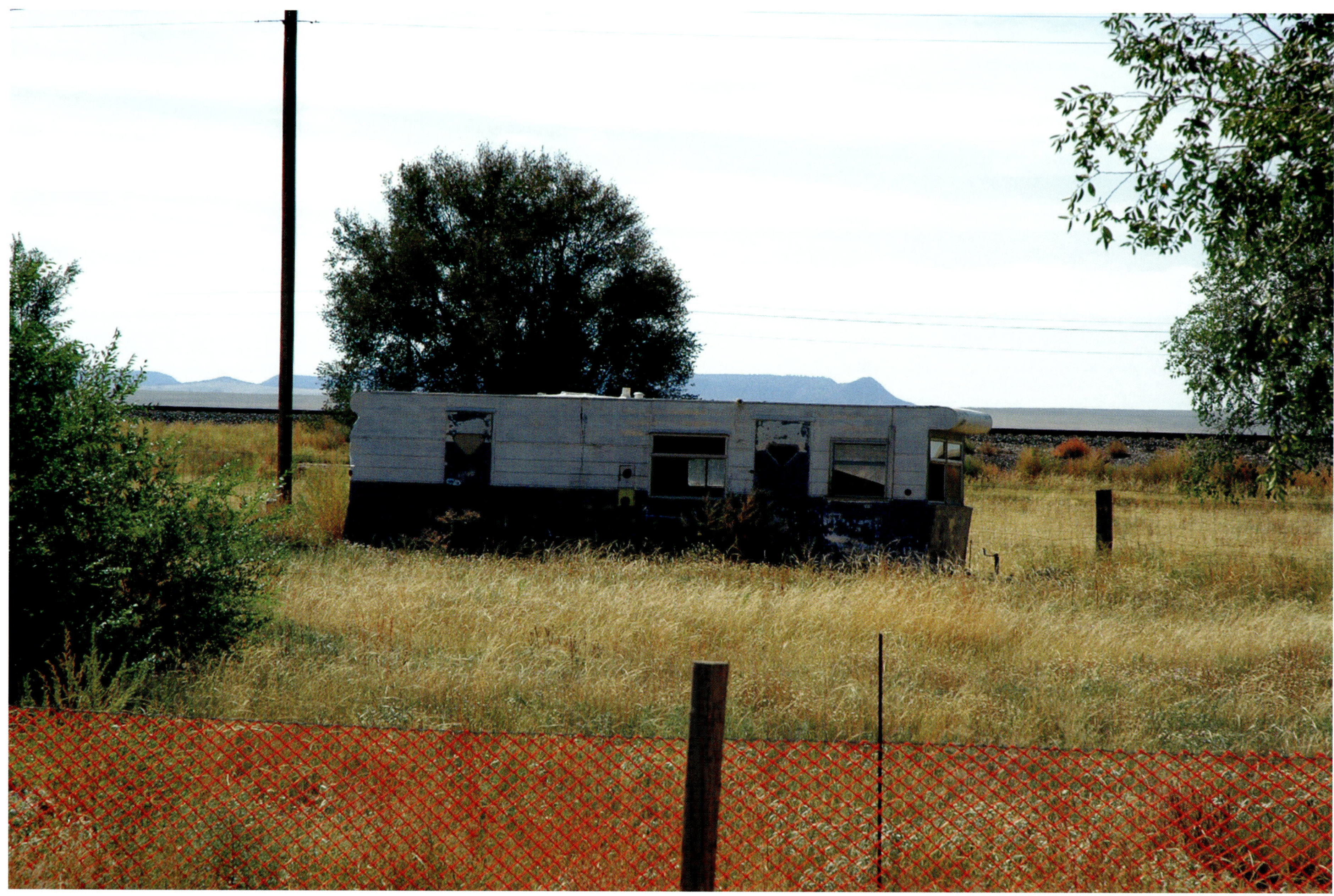

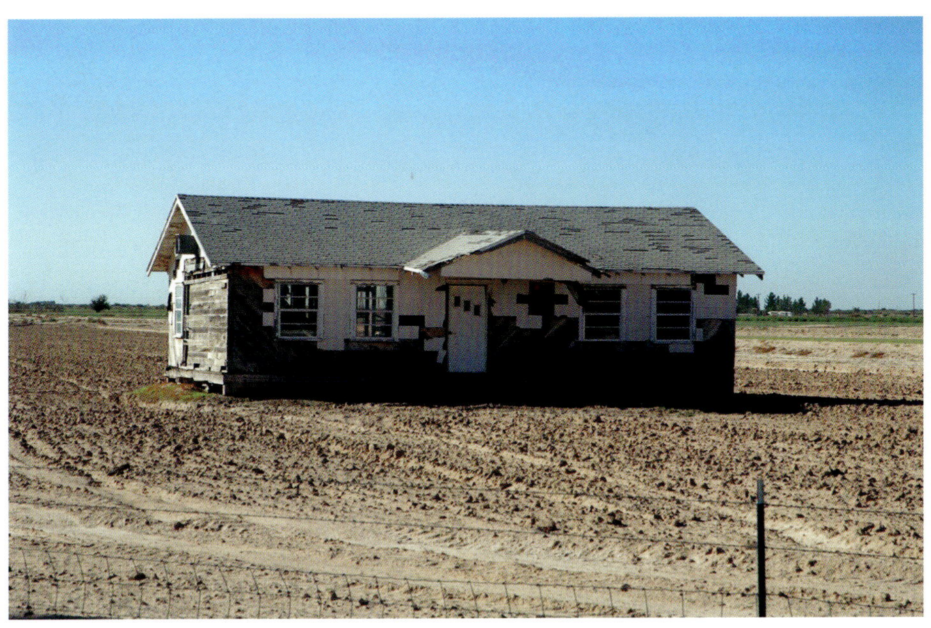

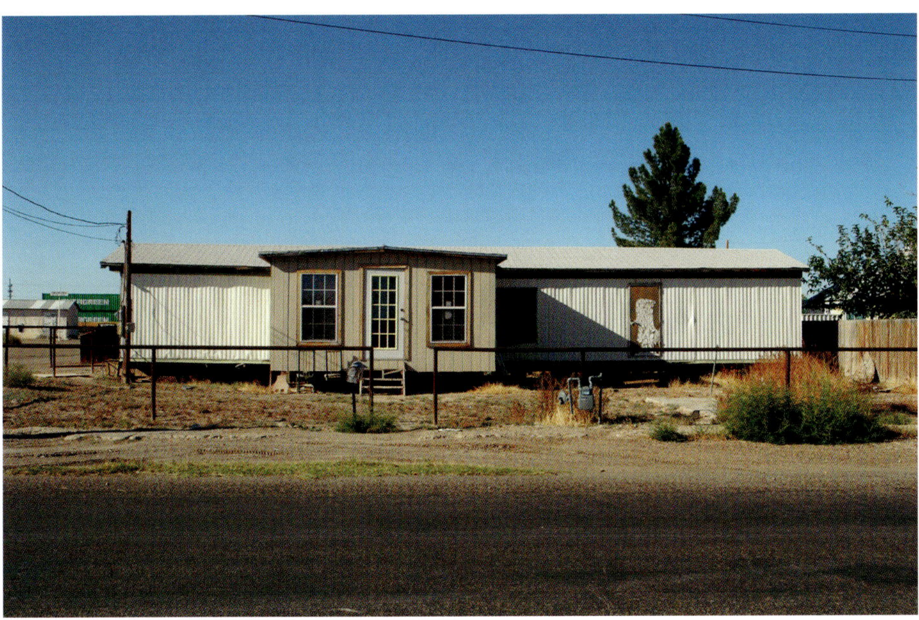

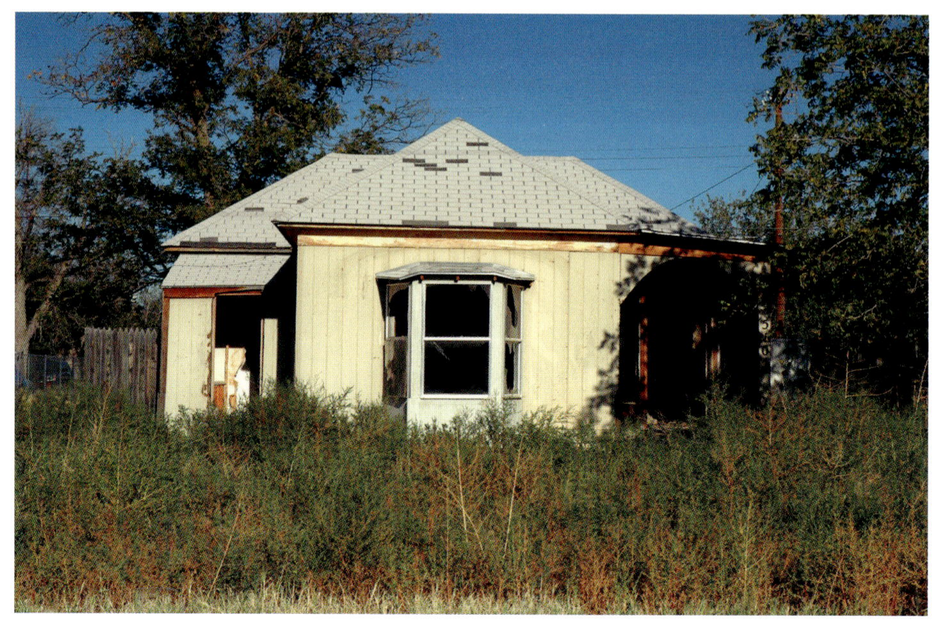

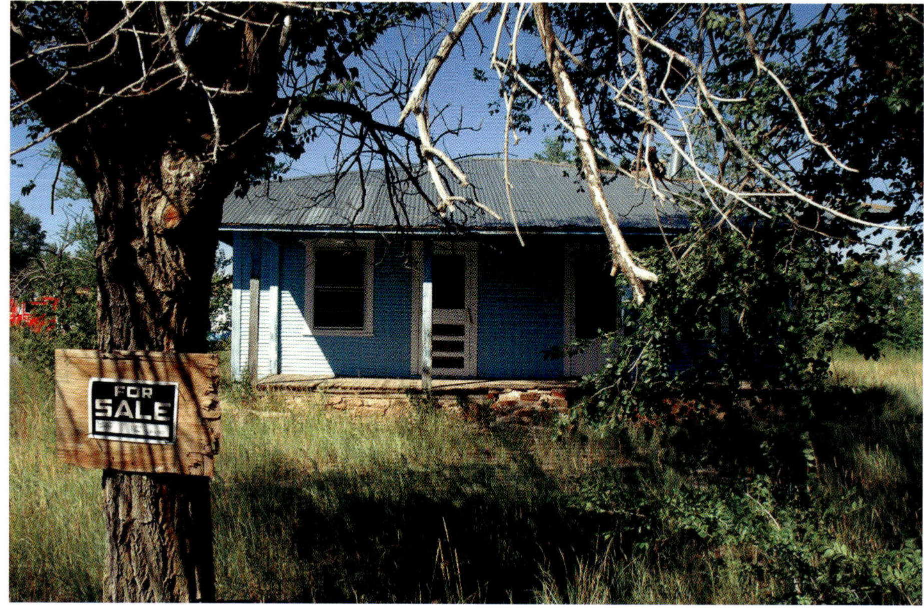

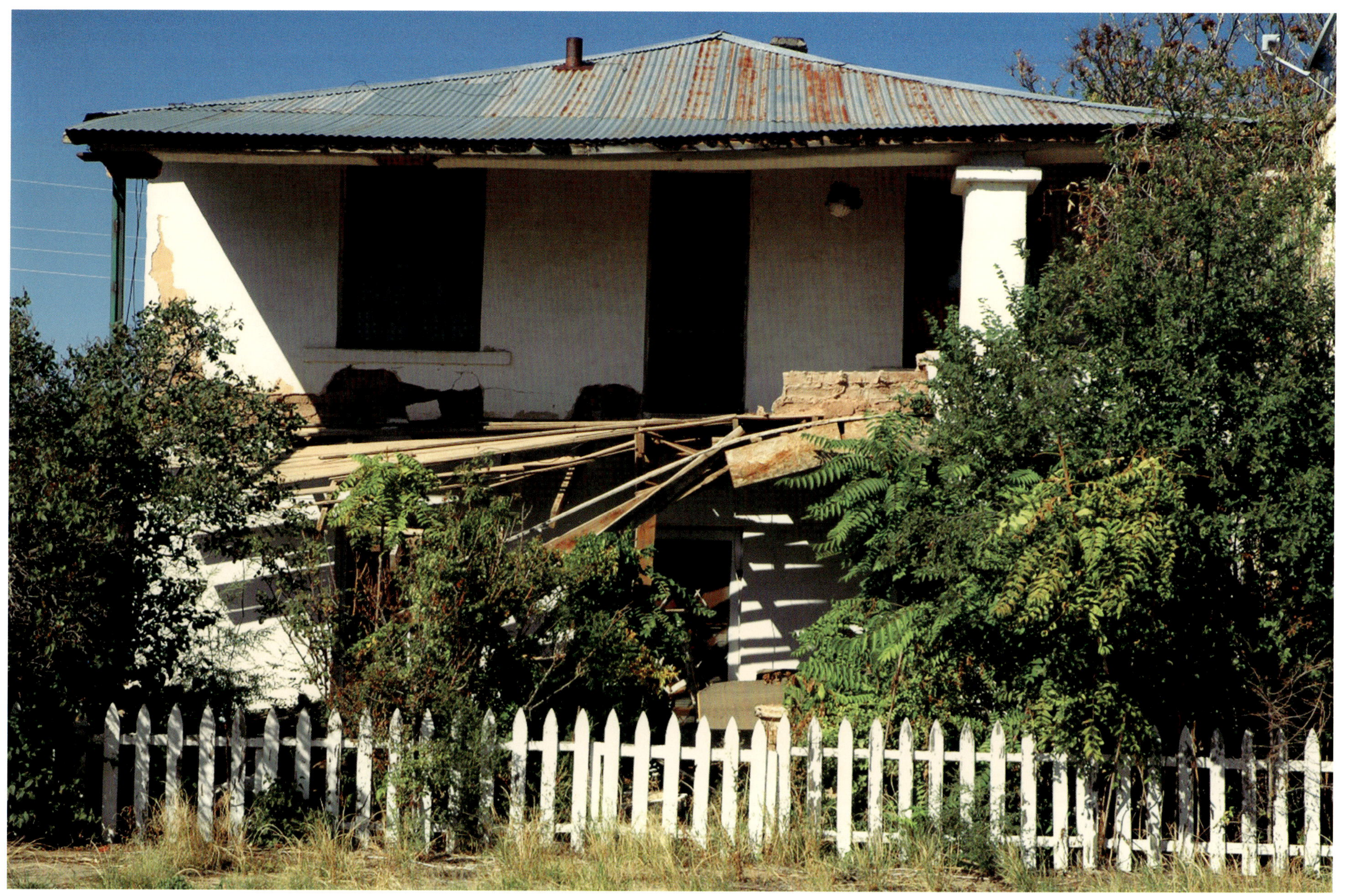

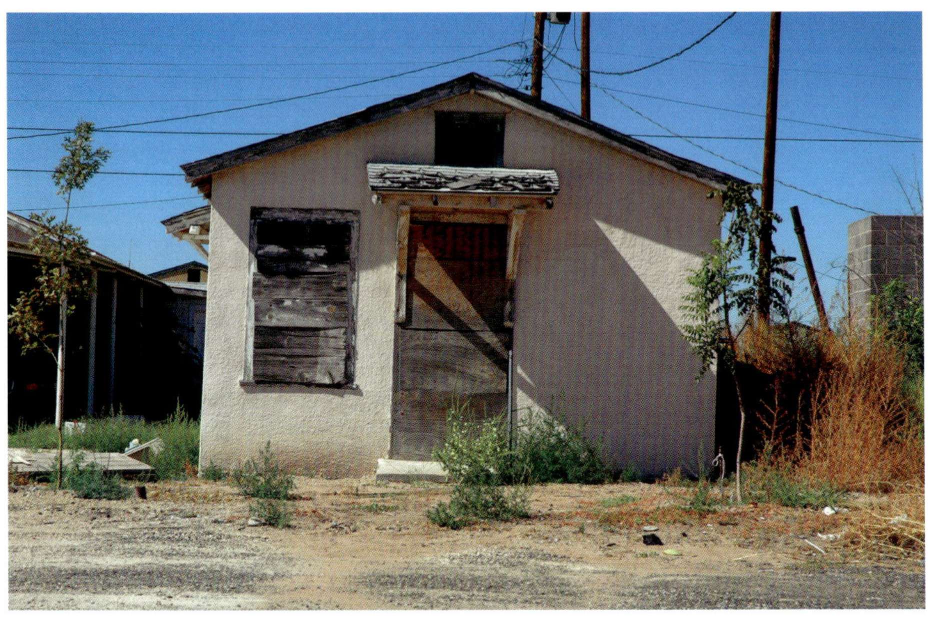
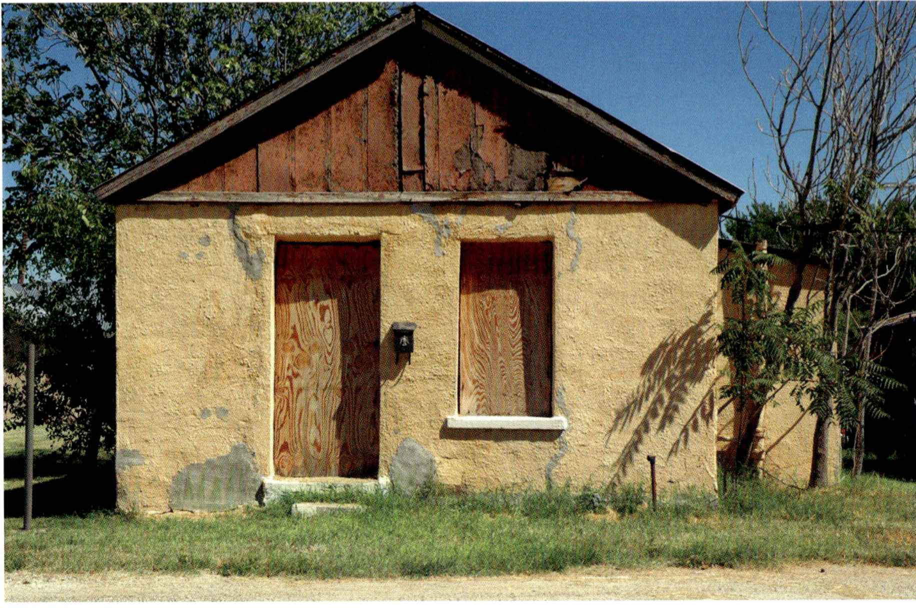
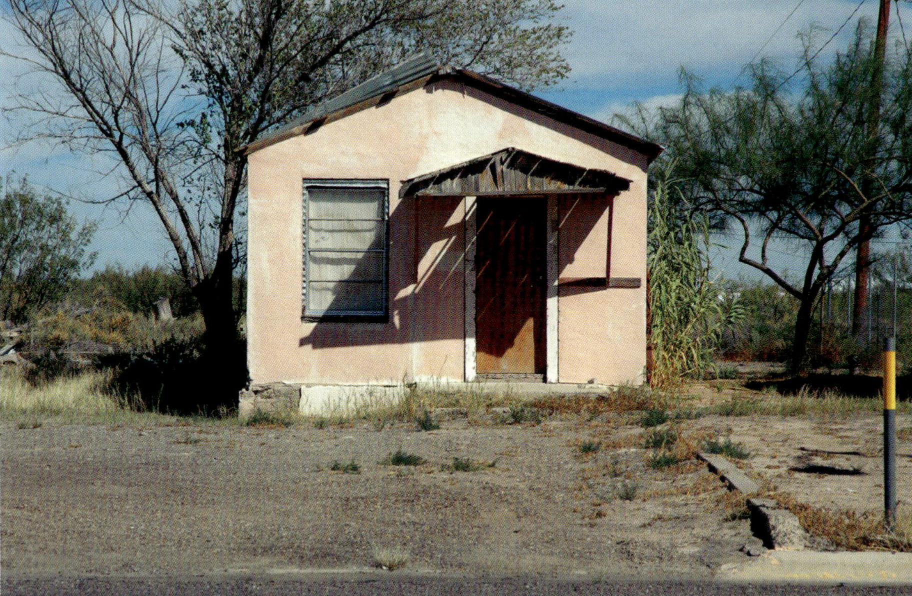
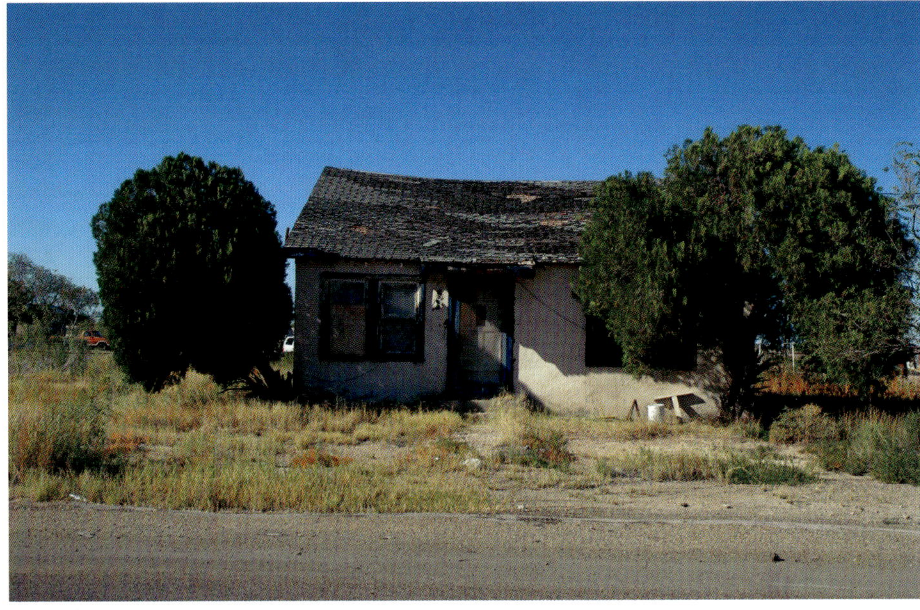

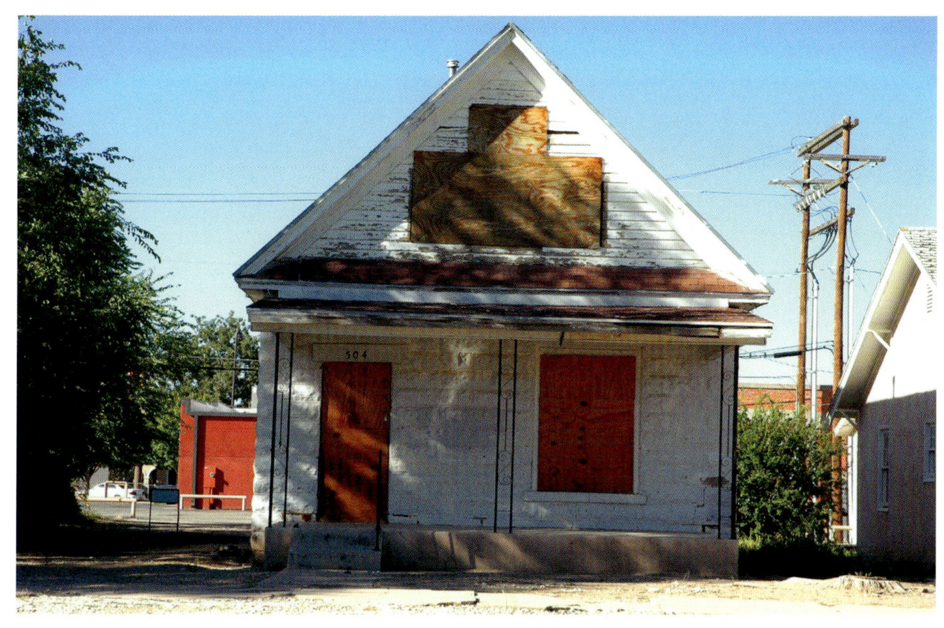
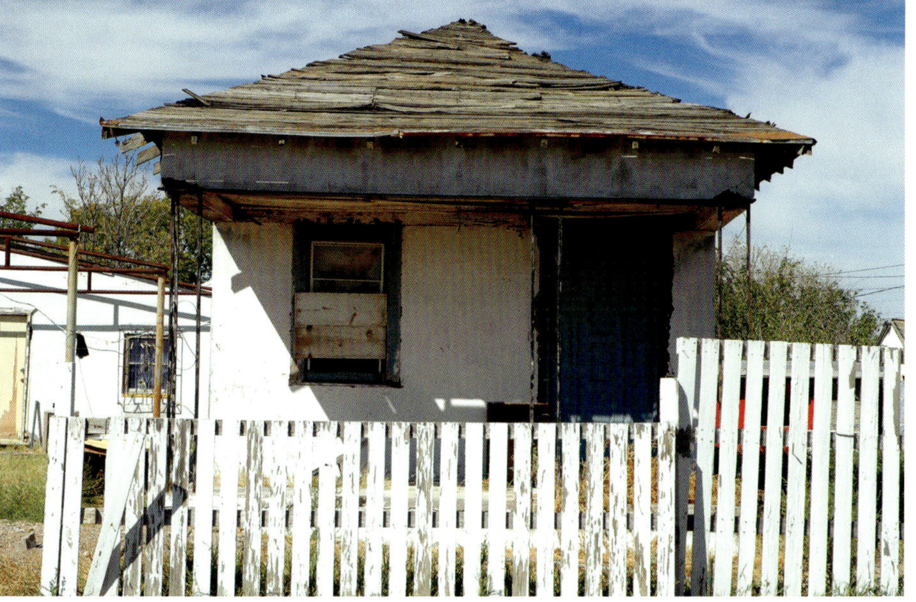
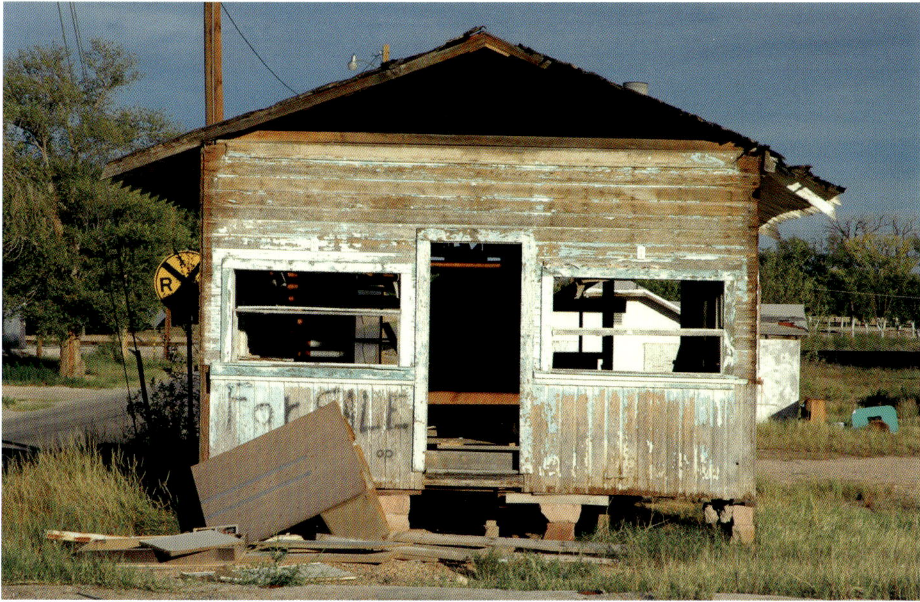
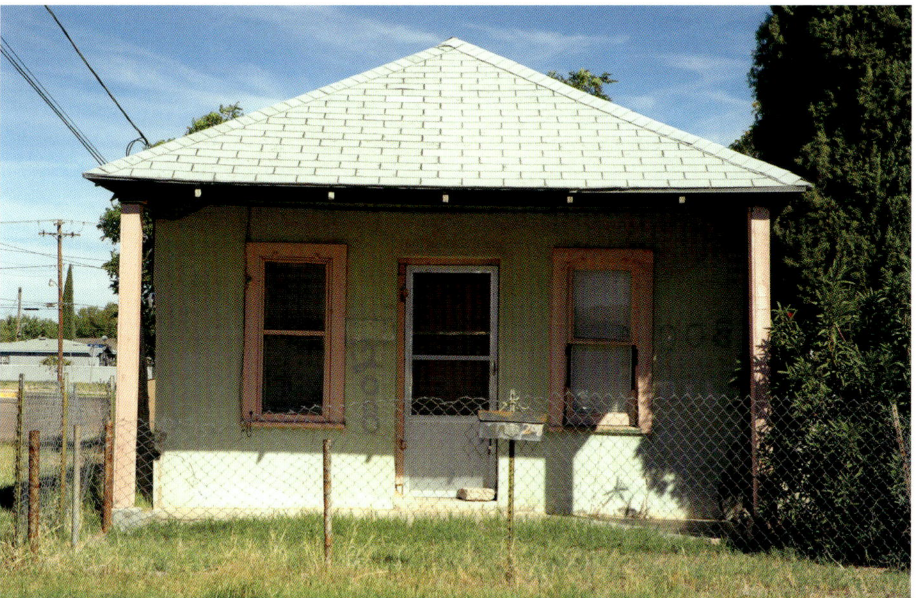

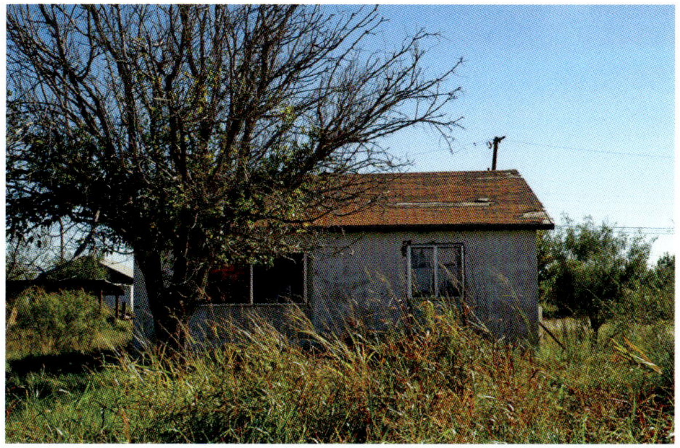

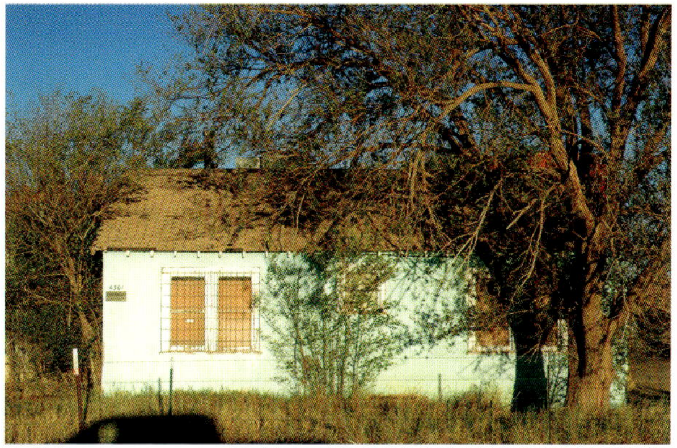

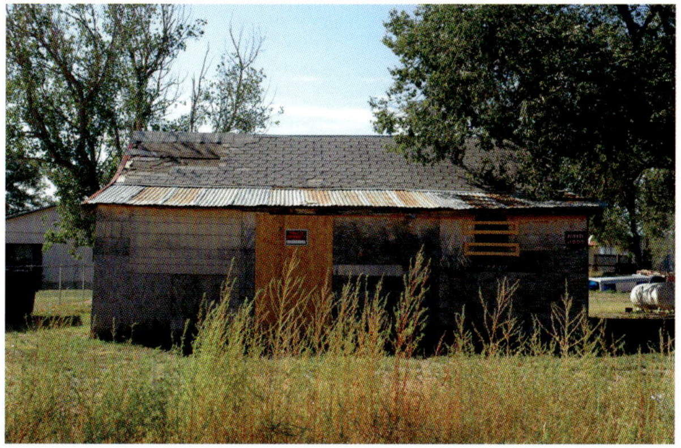

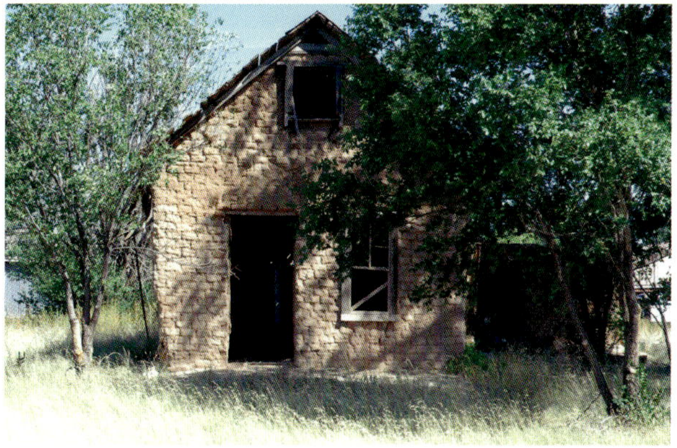

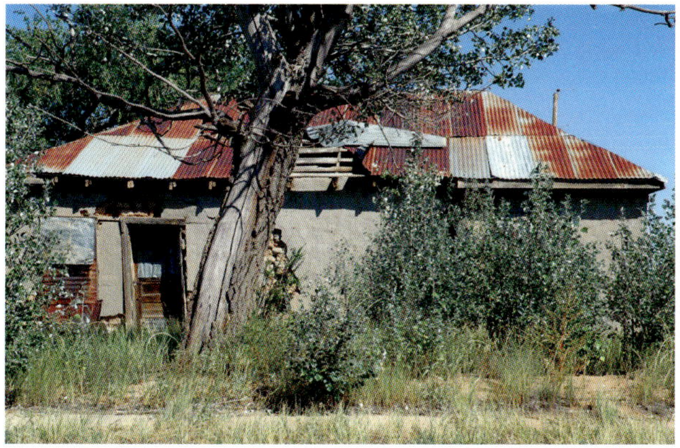

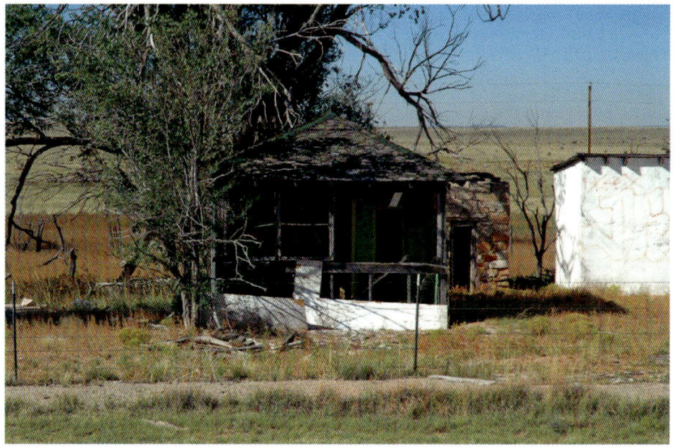

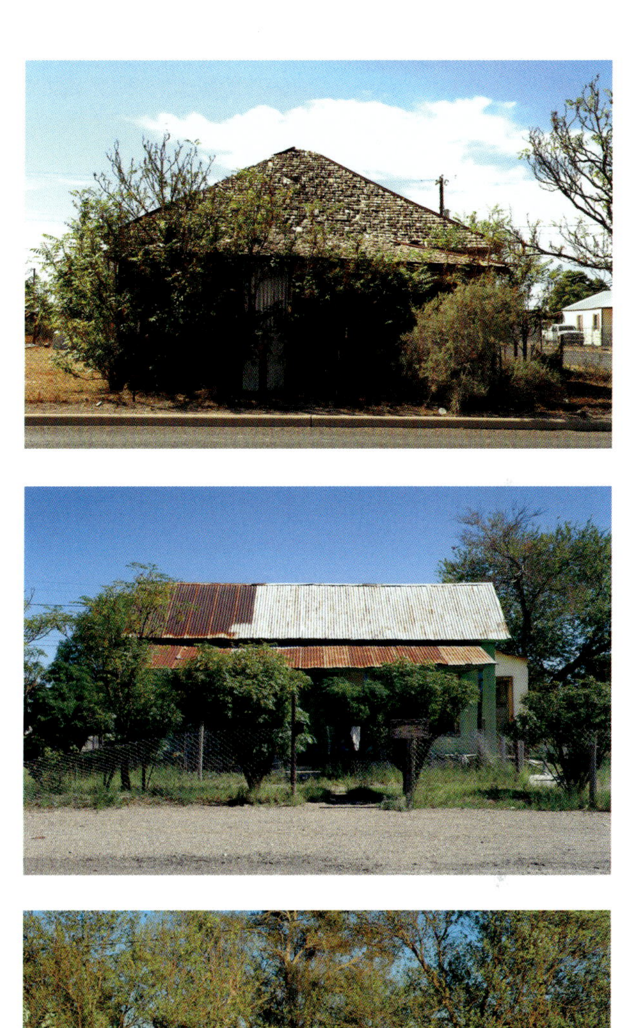
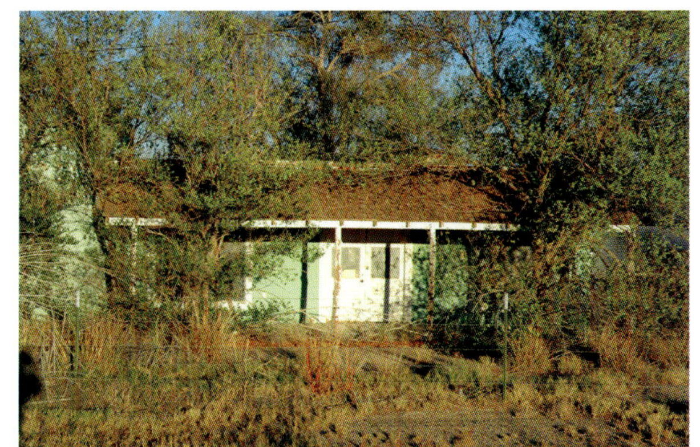

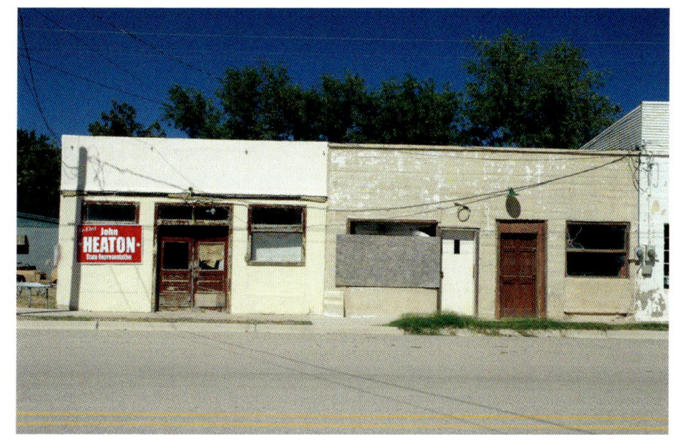
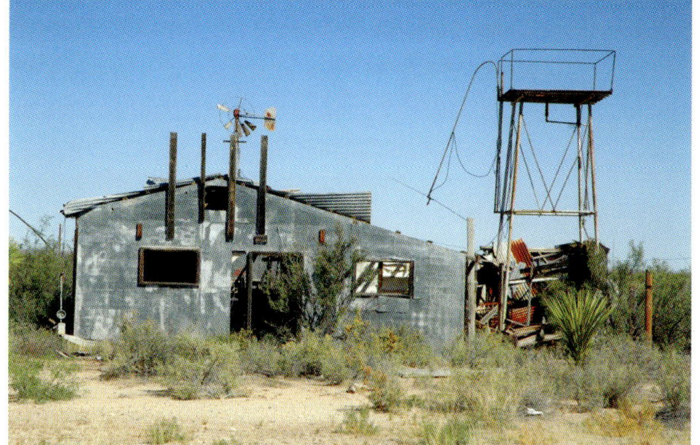
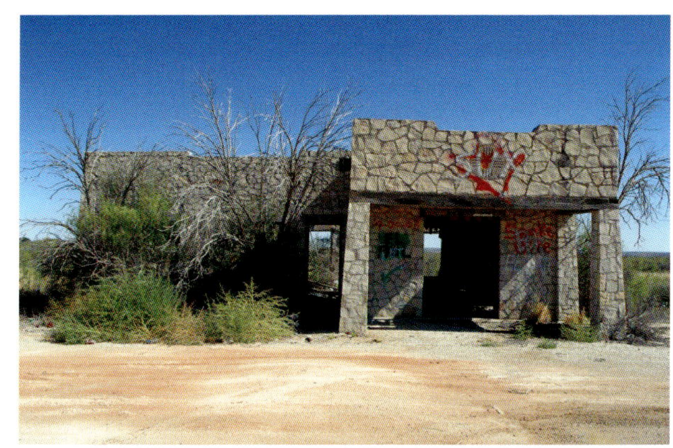
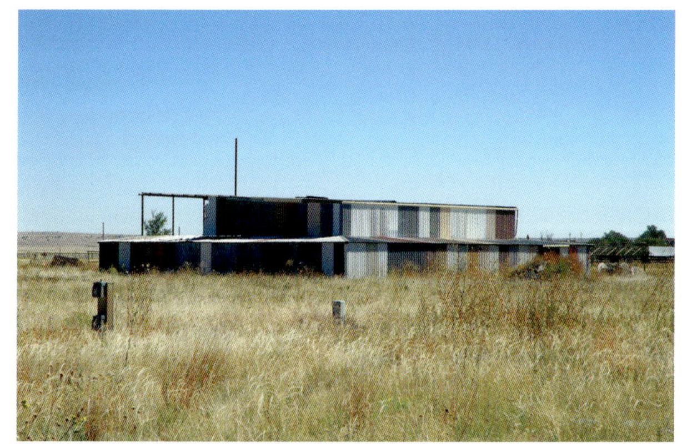
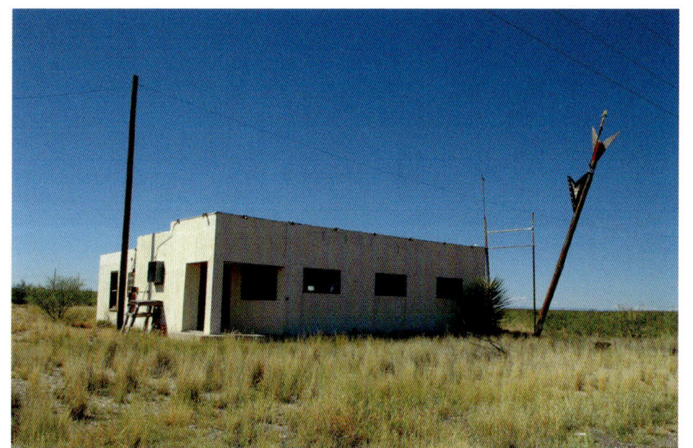
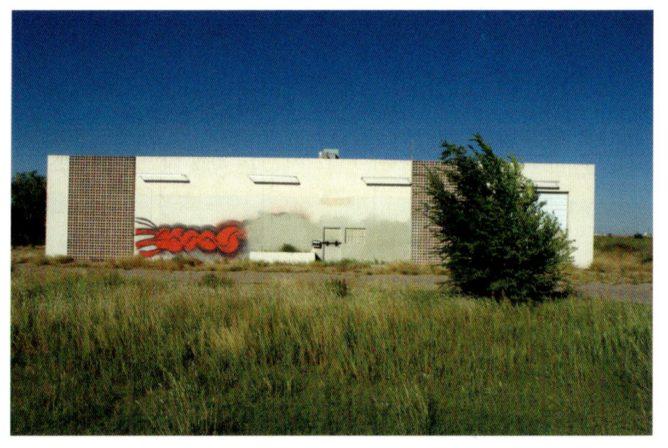
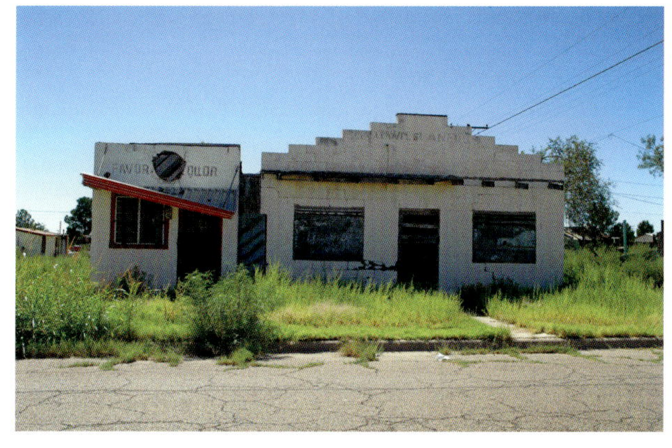
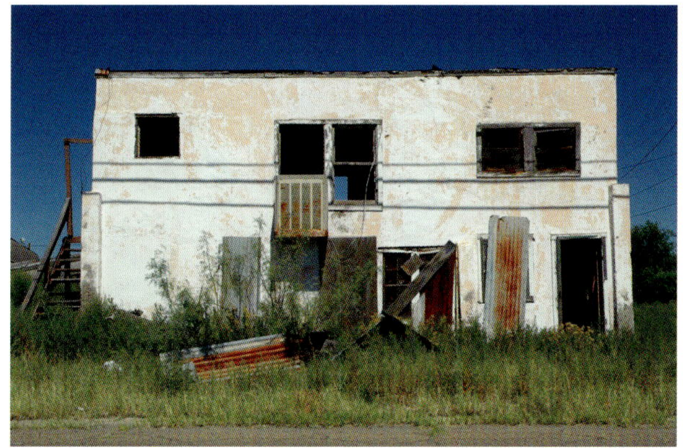
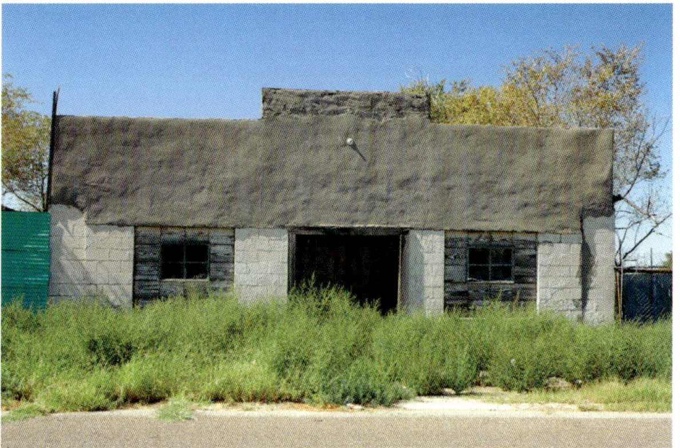

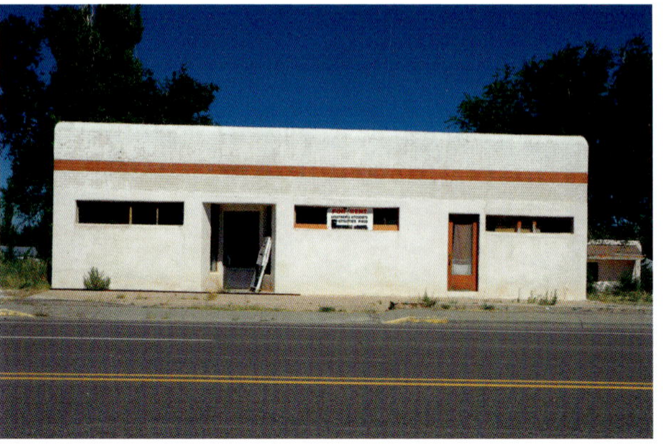

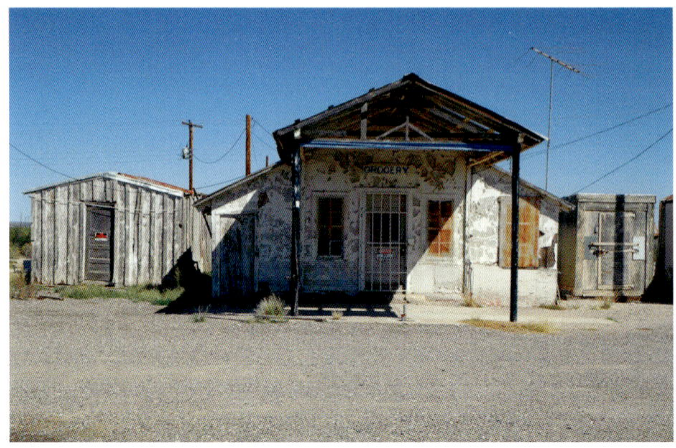

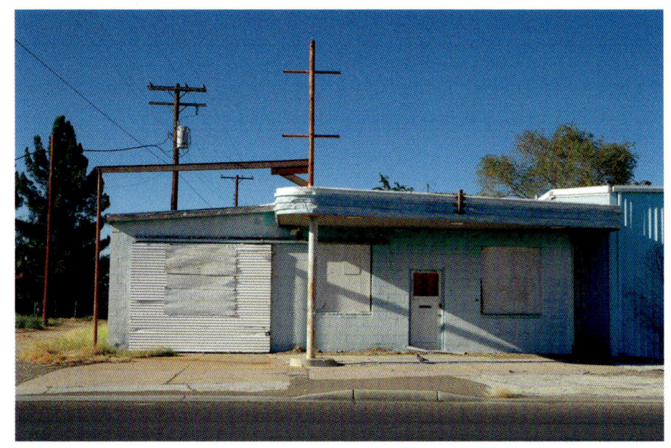

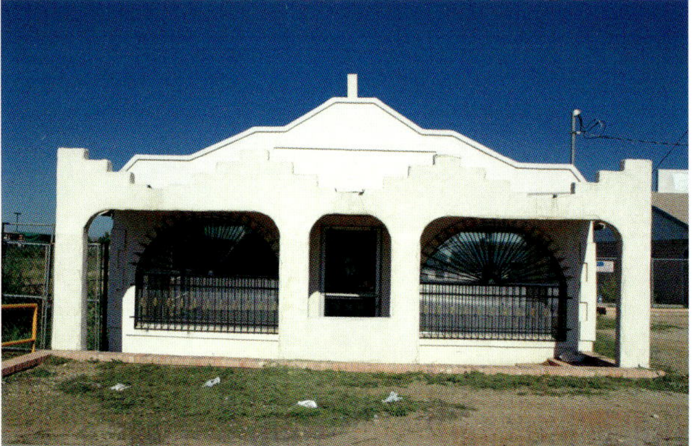

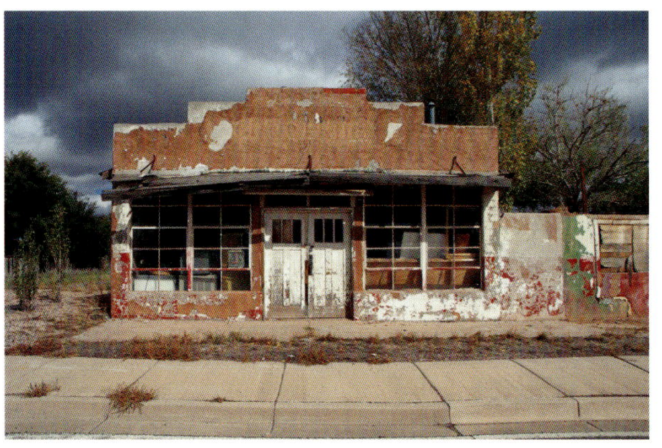

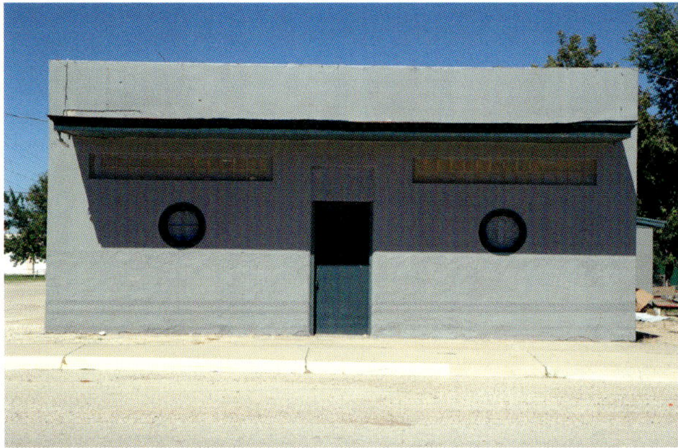

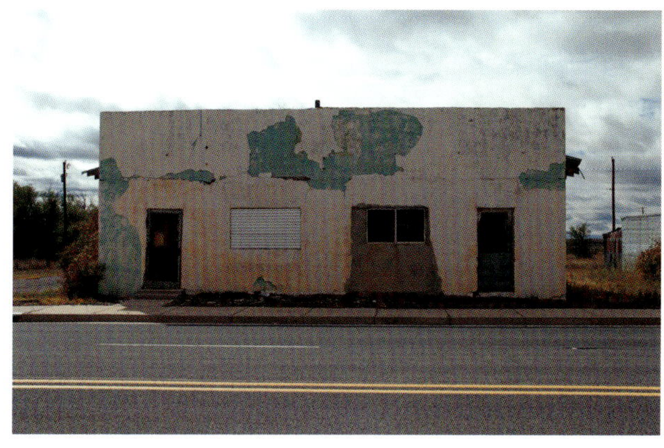 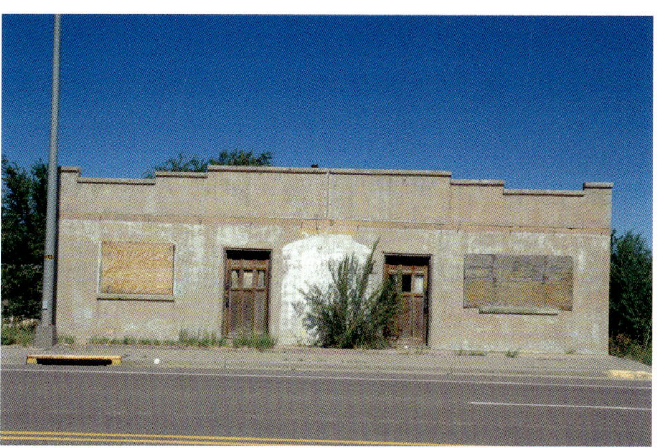

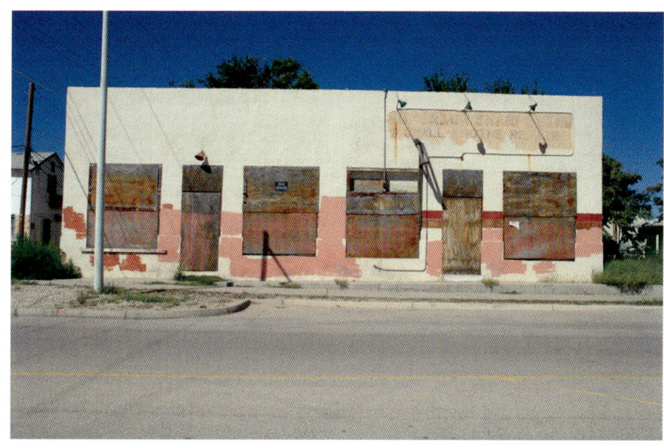 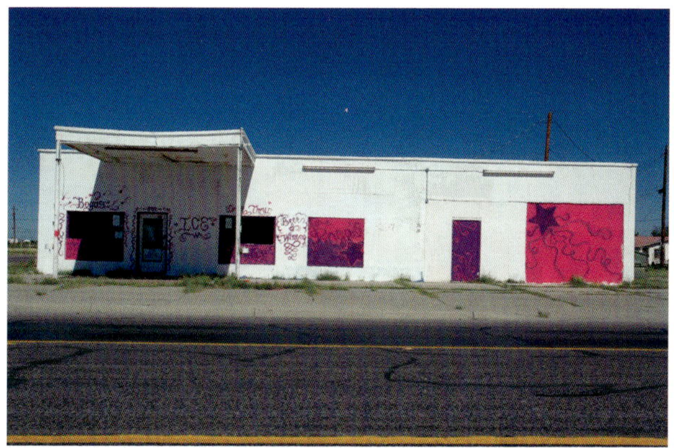

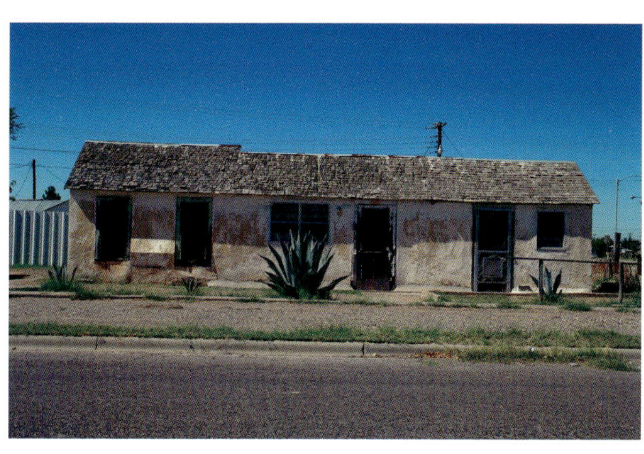 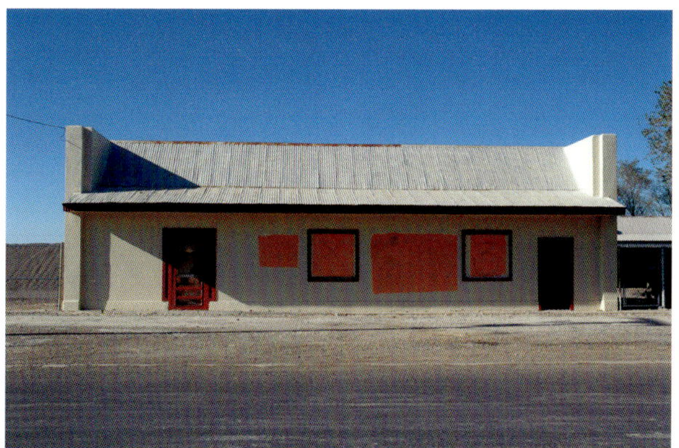

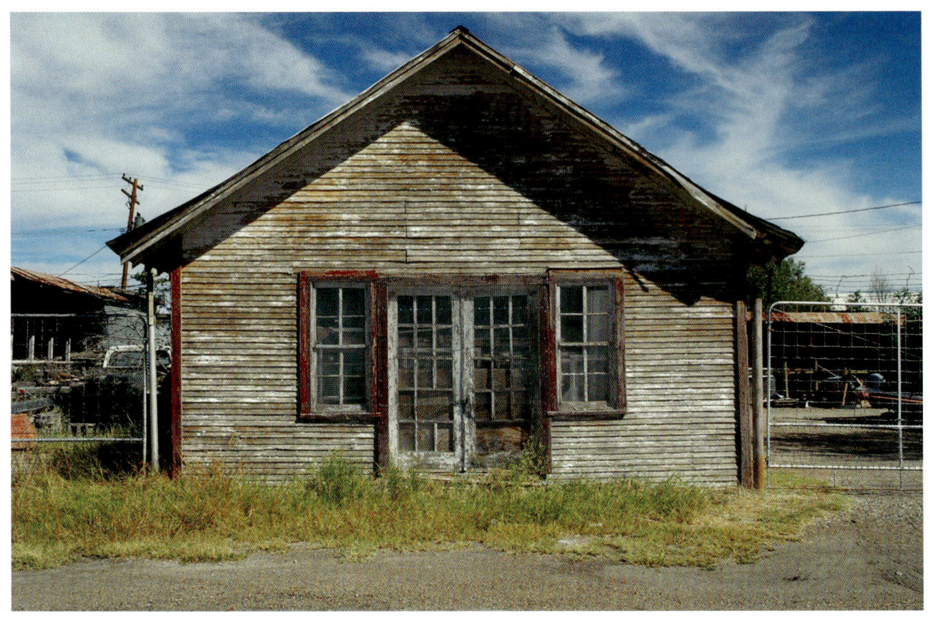
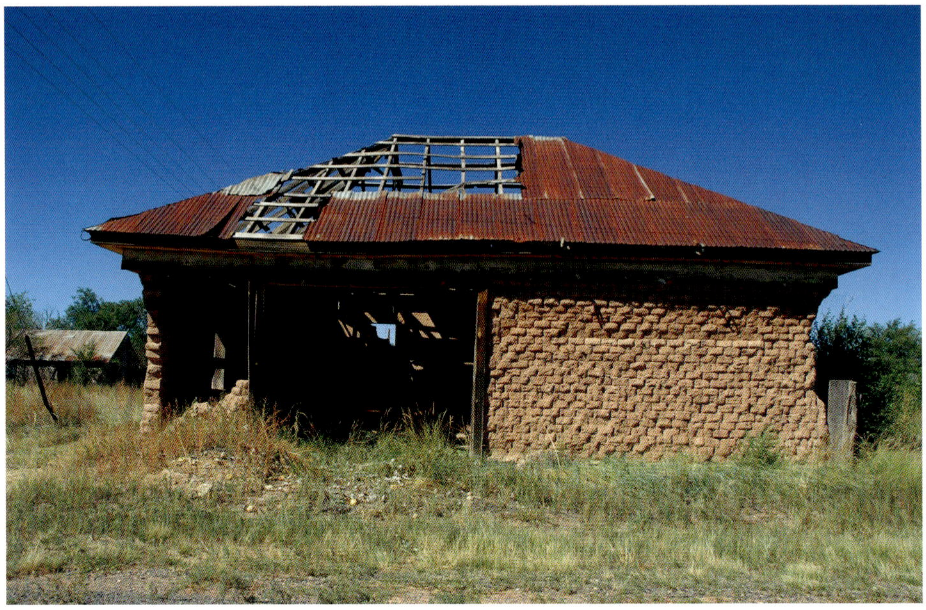
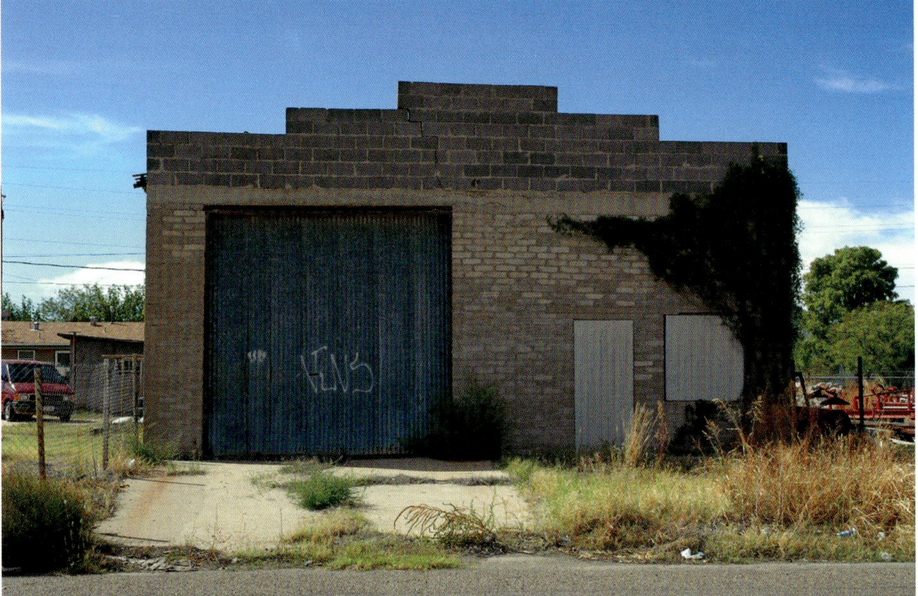
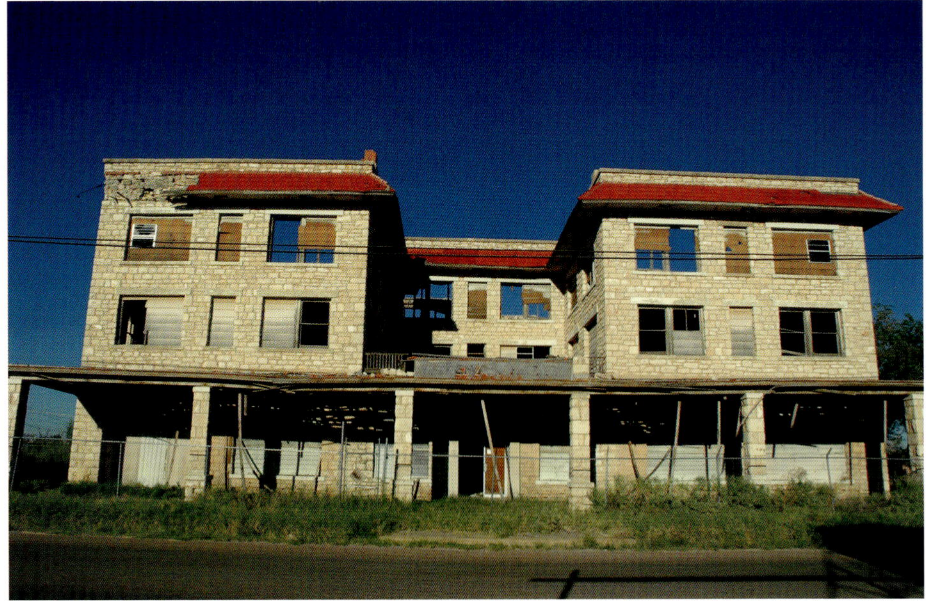

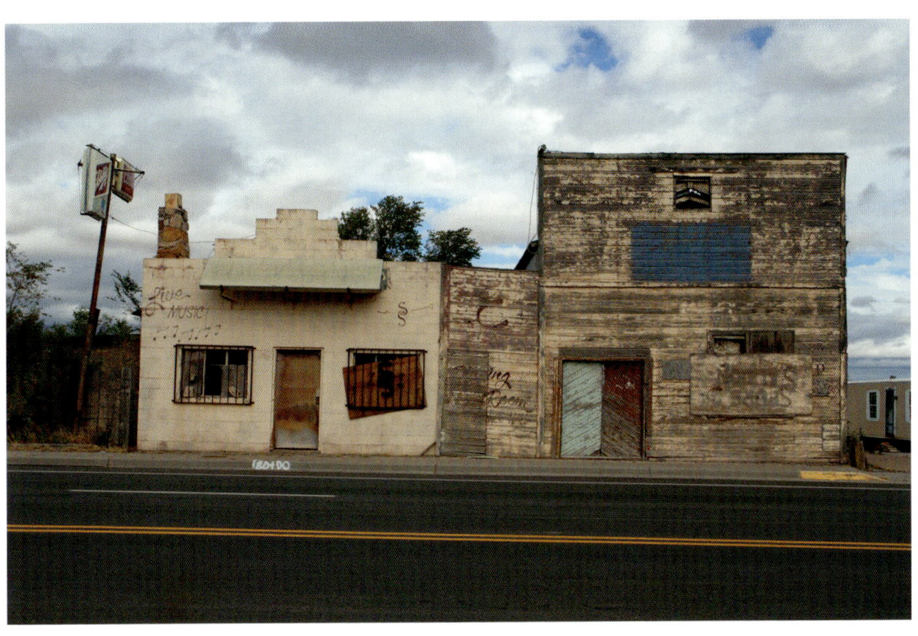

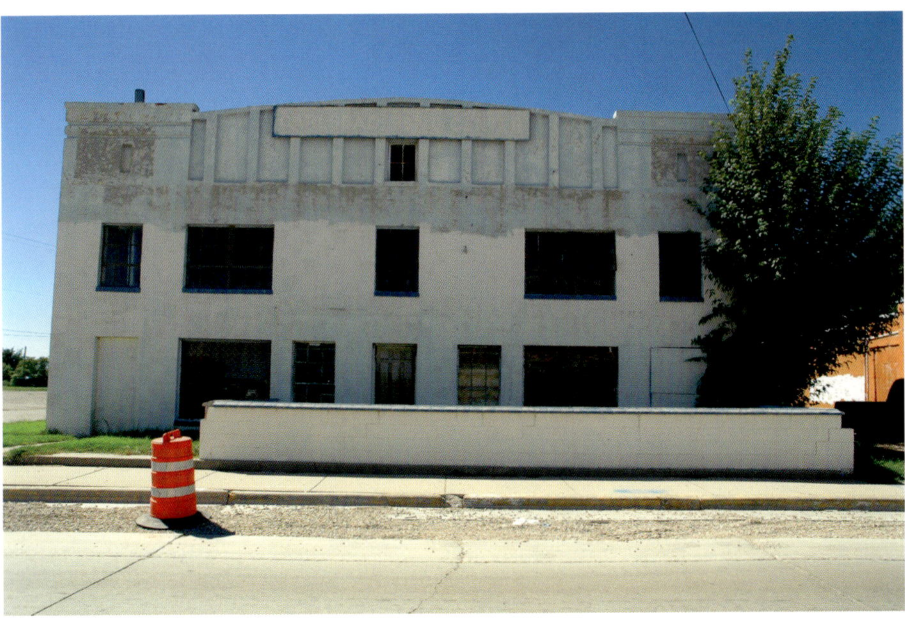

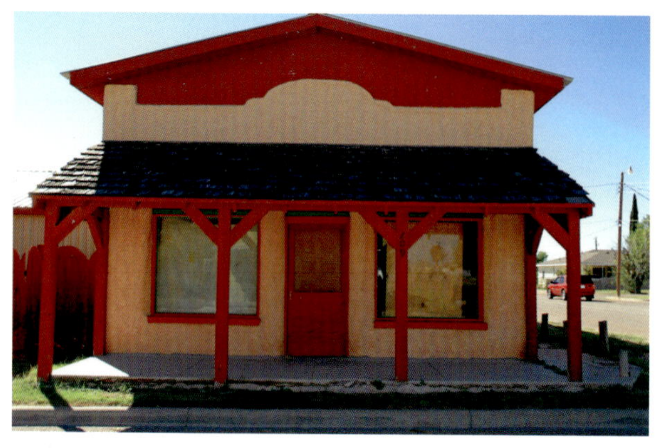
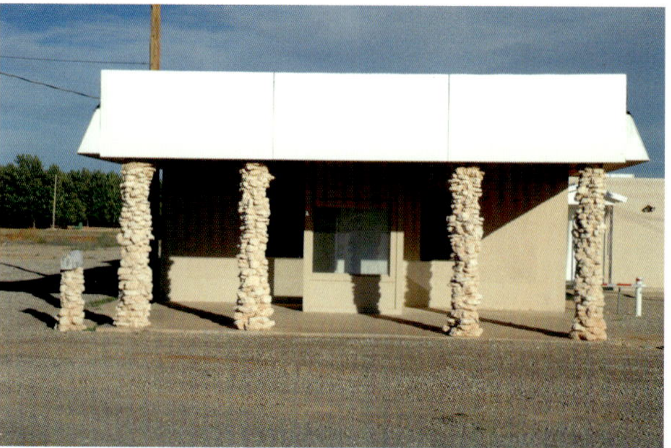
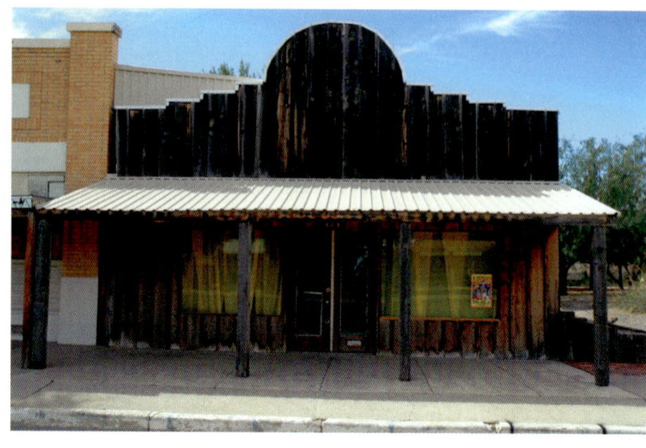
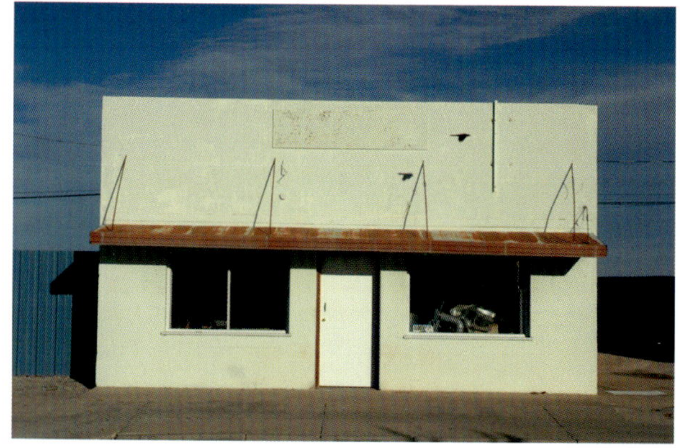
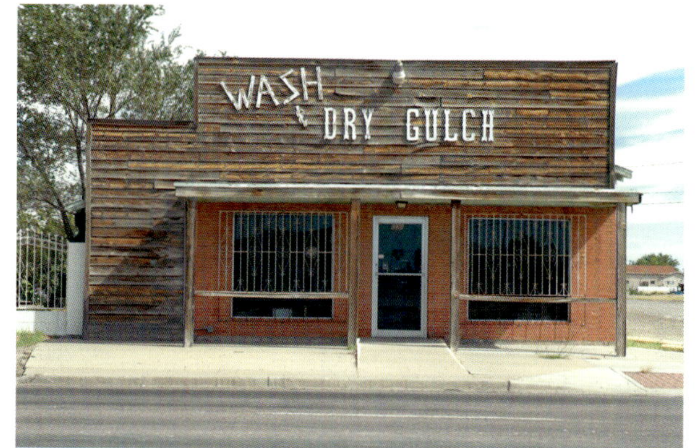

WASH
DRY GULCH

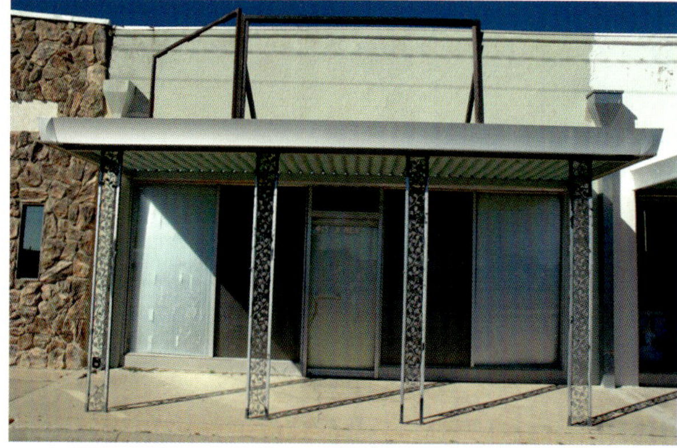
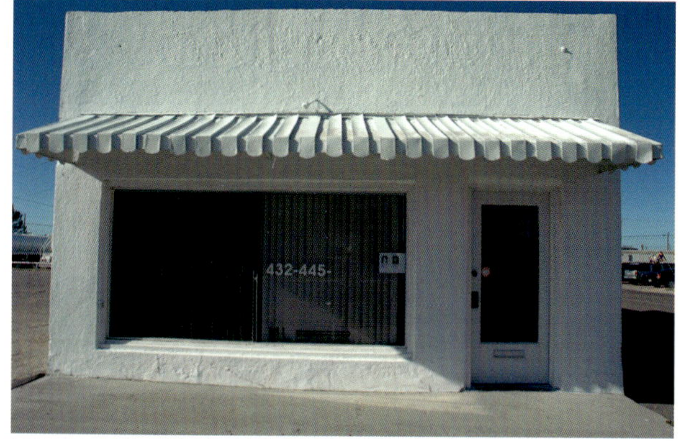
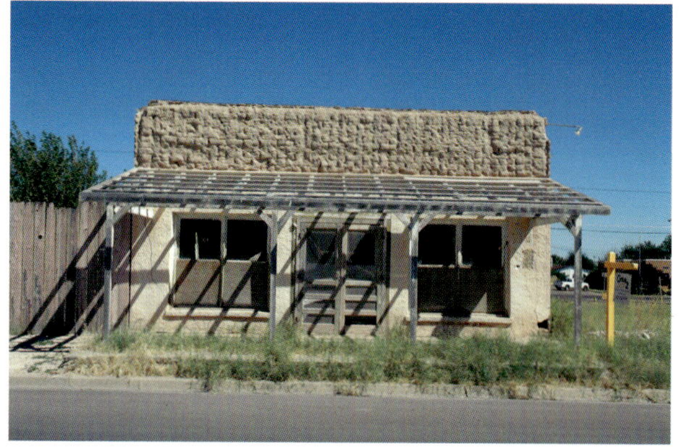
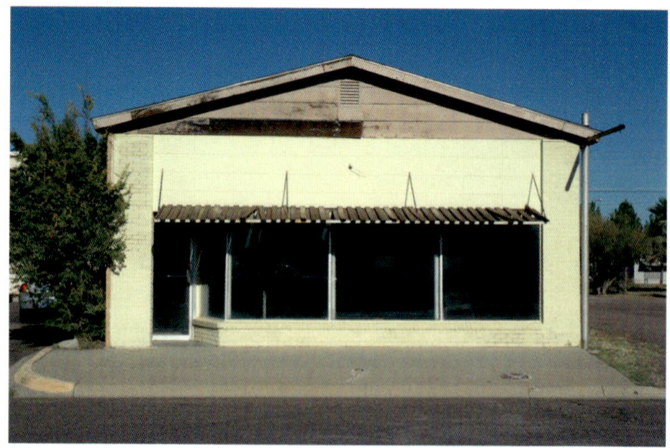

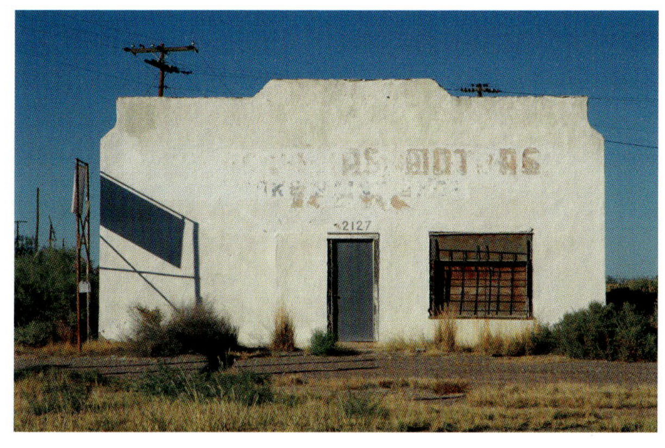
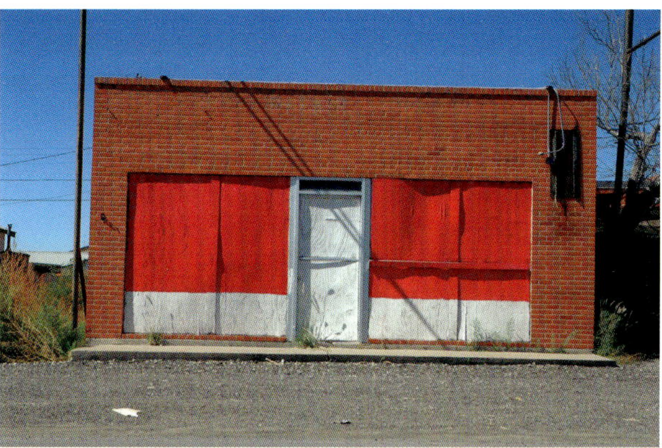
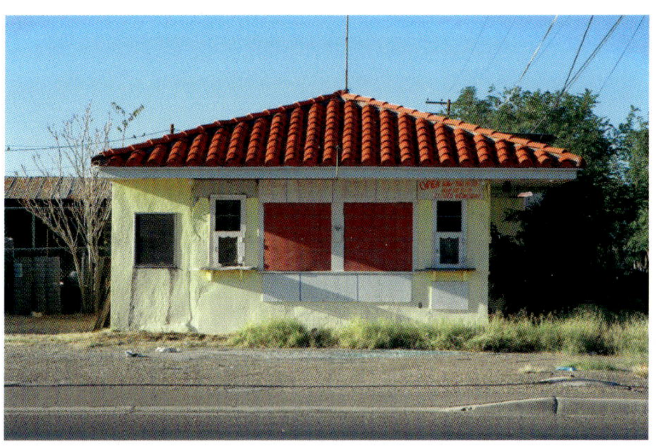
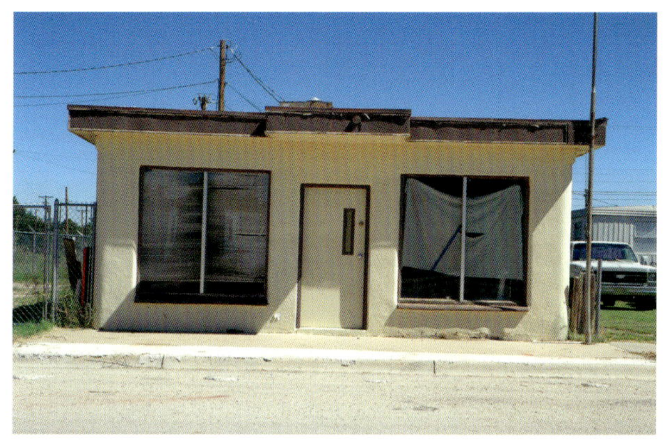
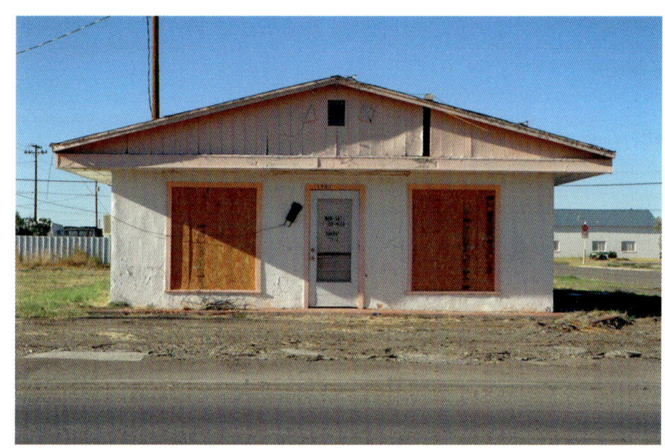
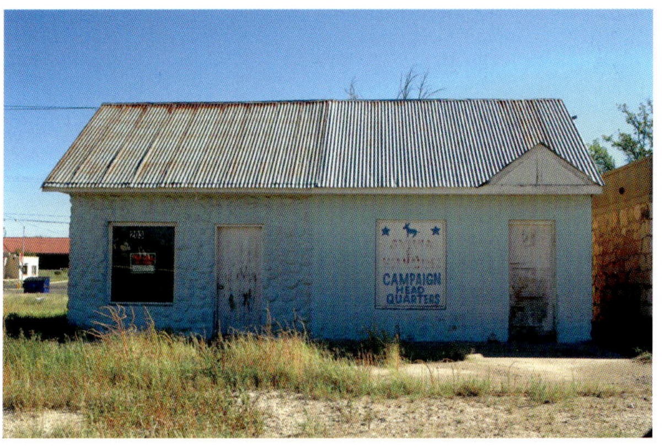
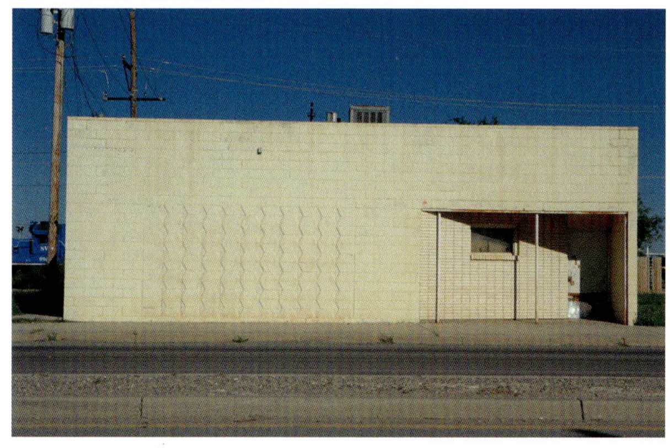
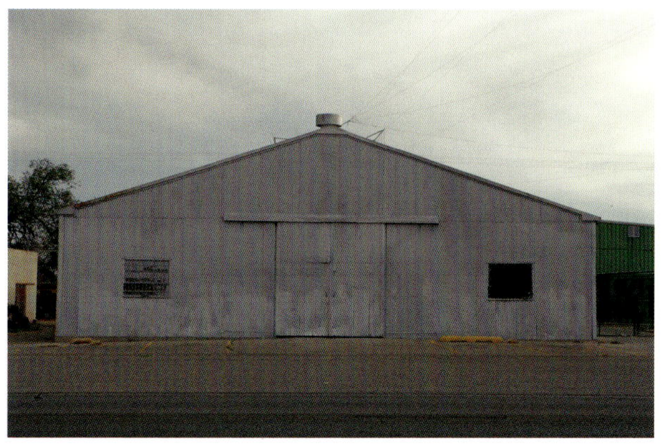
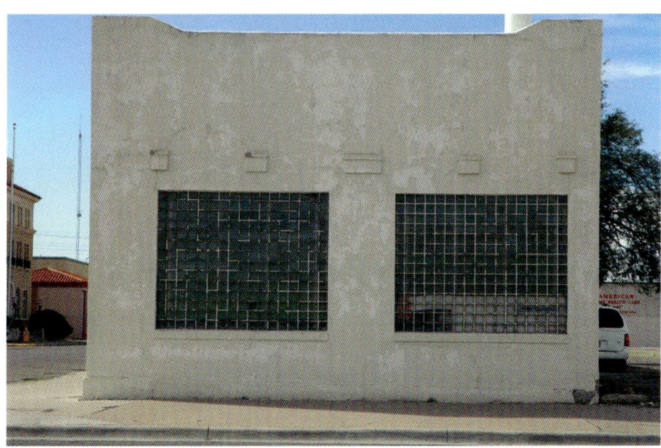

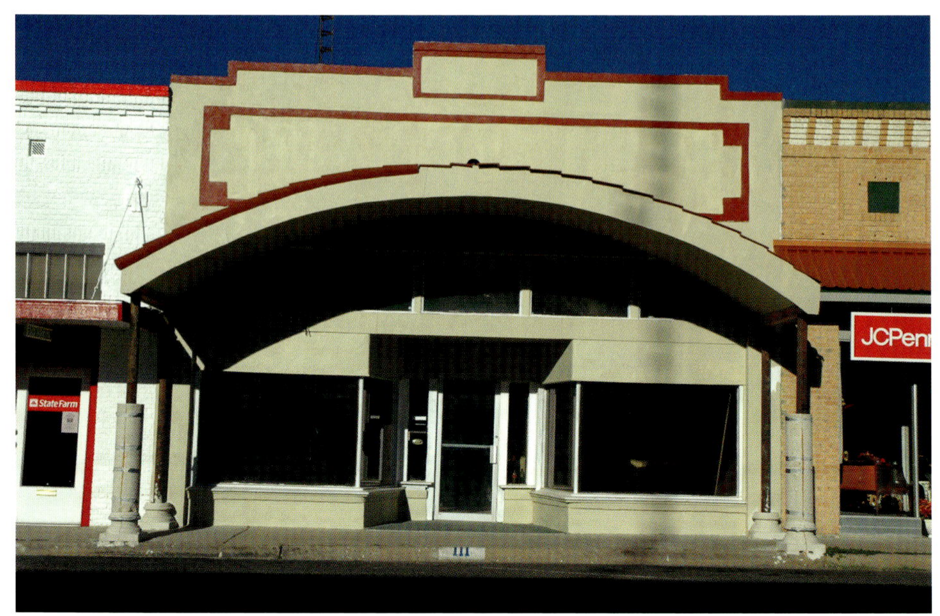

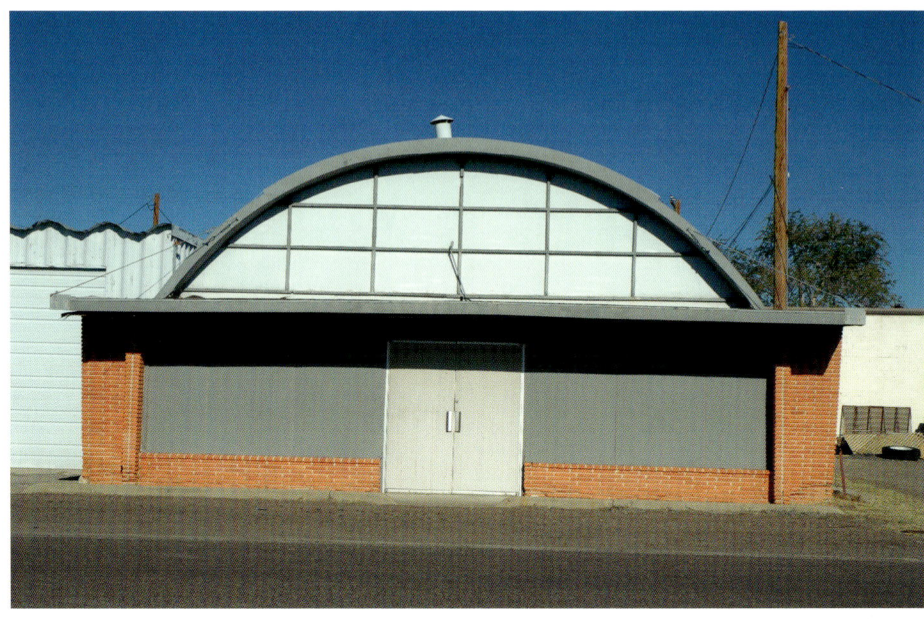

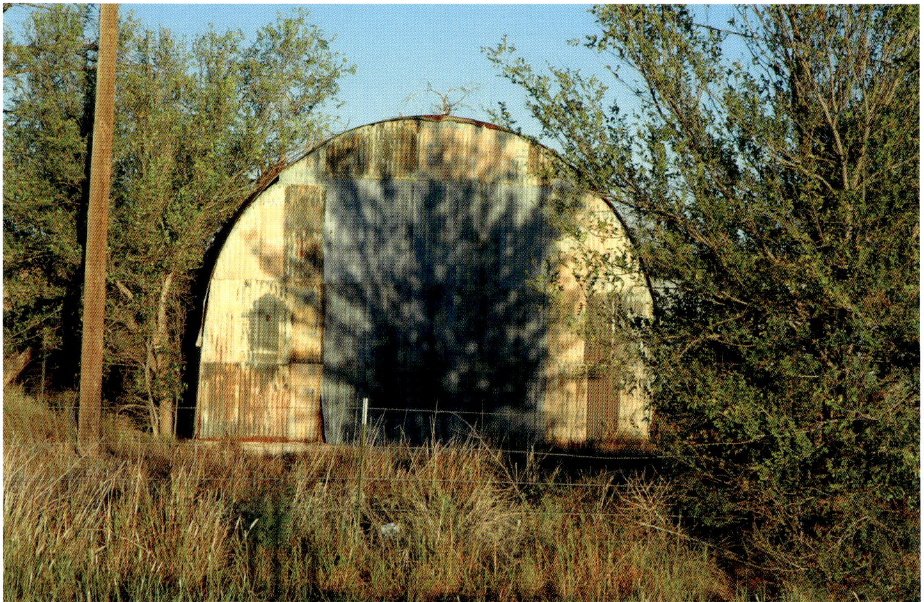

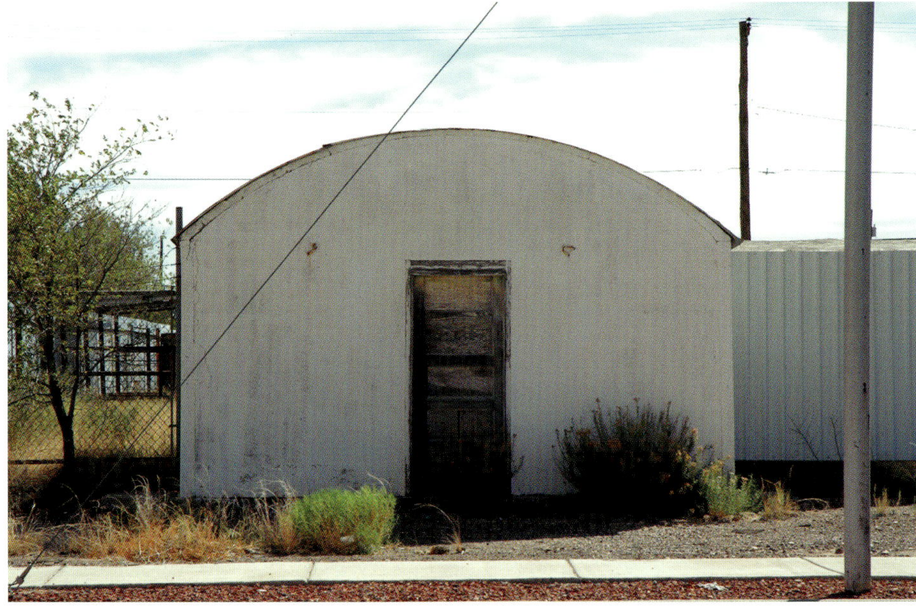

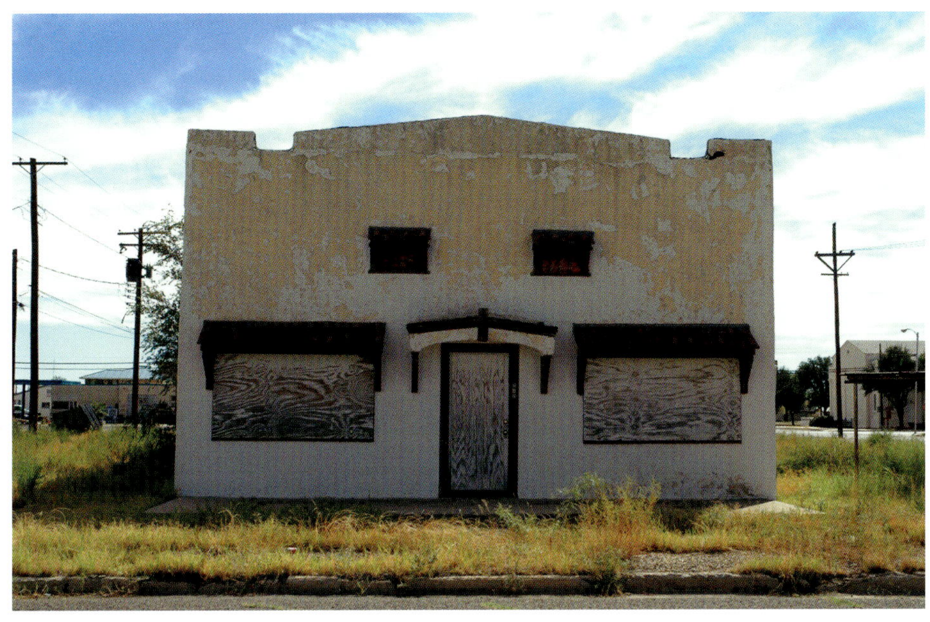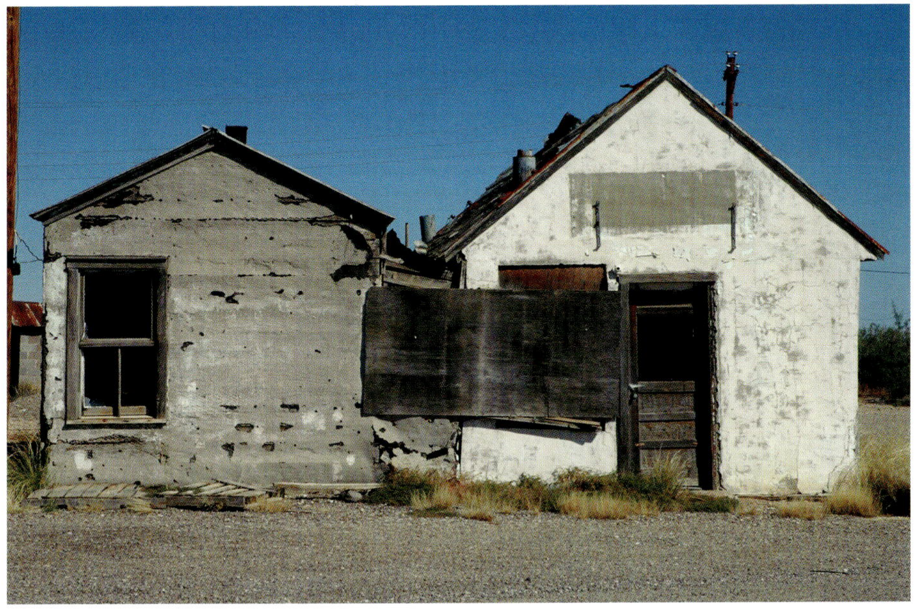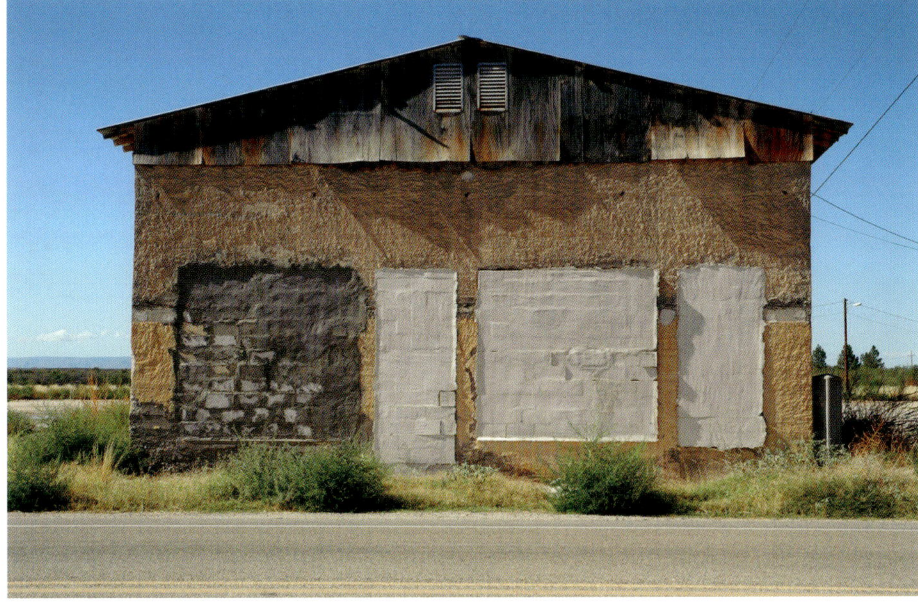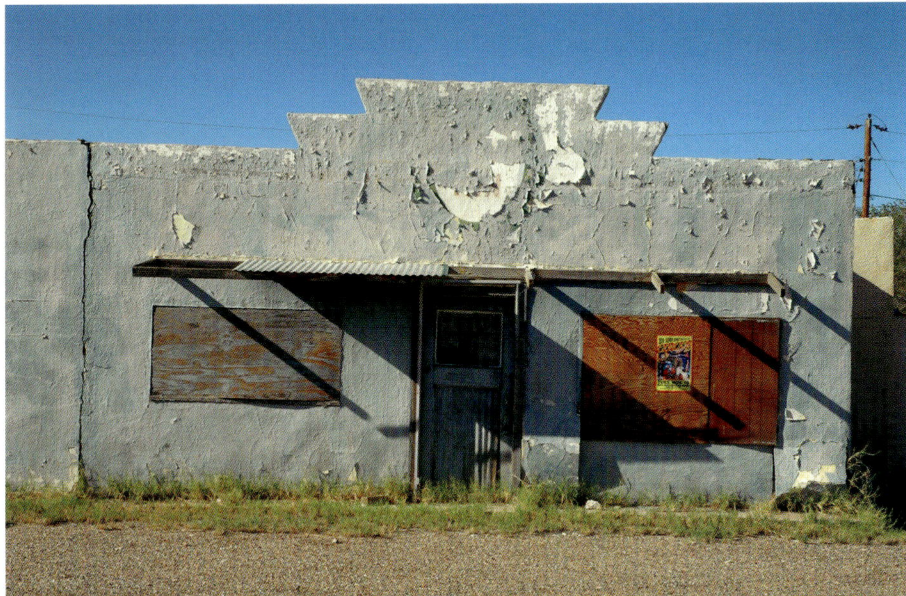

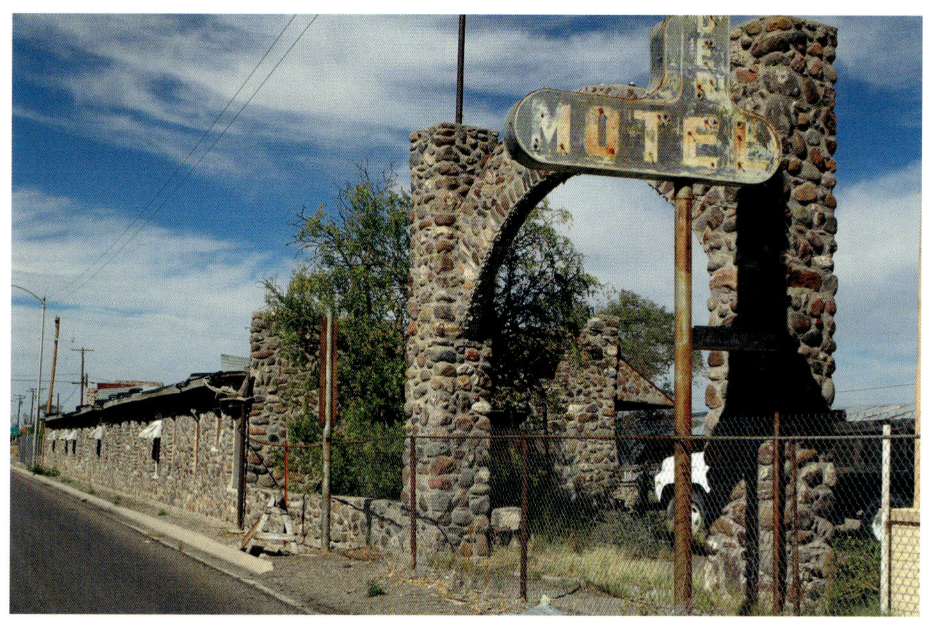
MOTEL

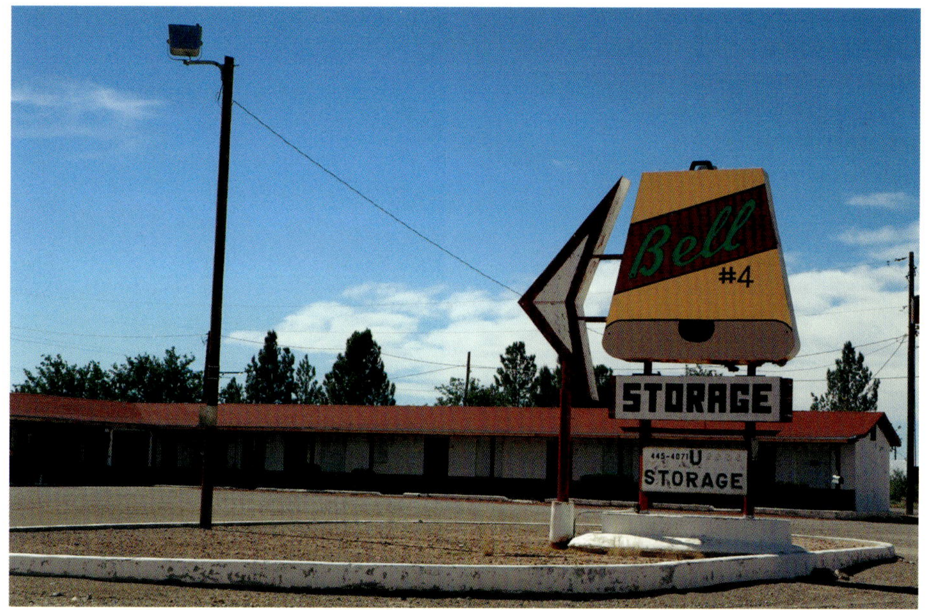
Bell
#4
STORAGE
445-4071
U
STORAGE

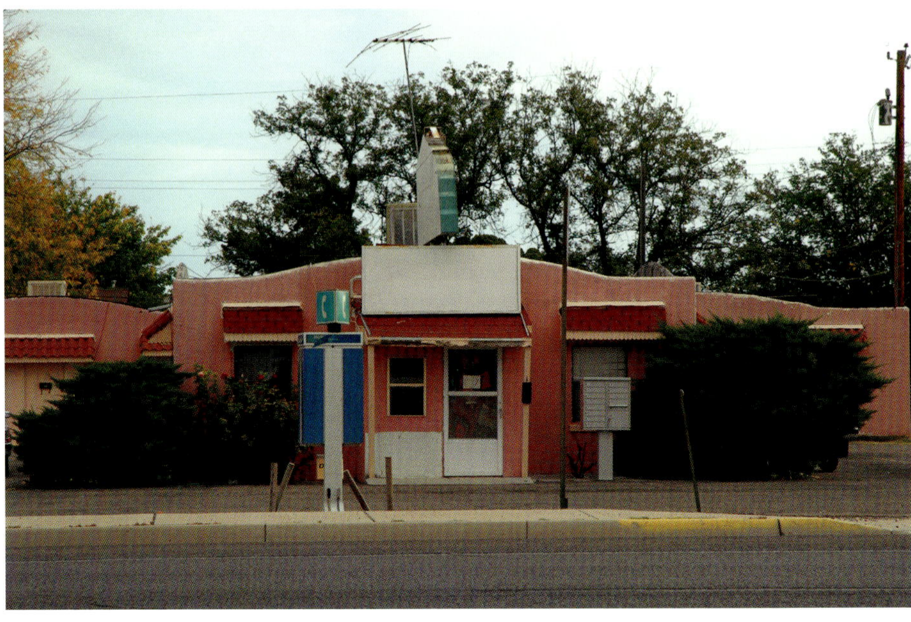

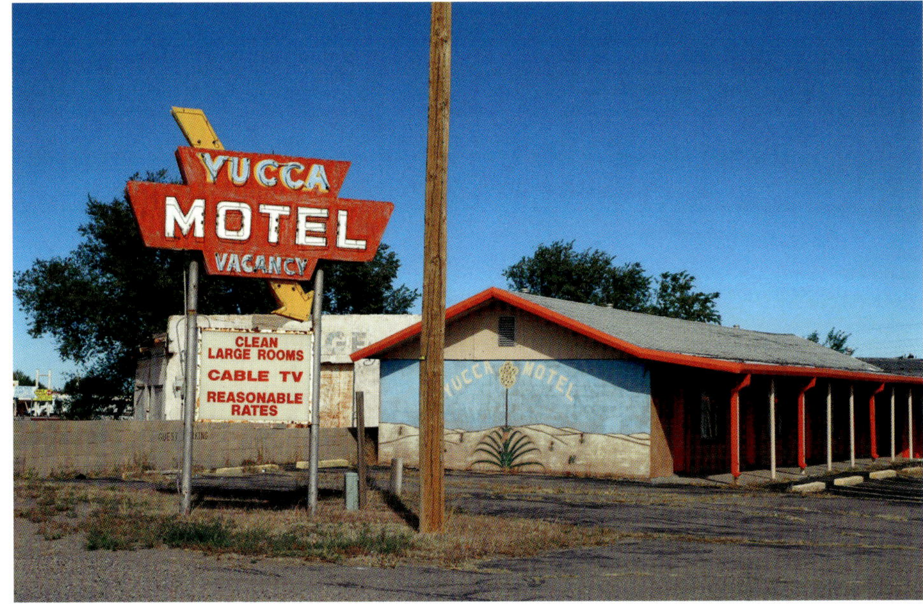
YUCCA
MOTEL
VACANCY
CLEAN
LARGE ROOMS
CABLE TV
REASONABLE
RATES
YUCCA MOTEL

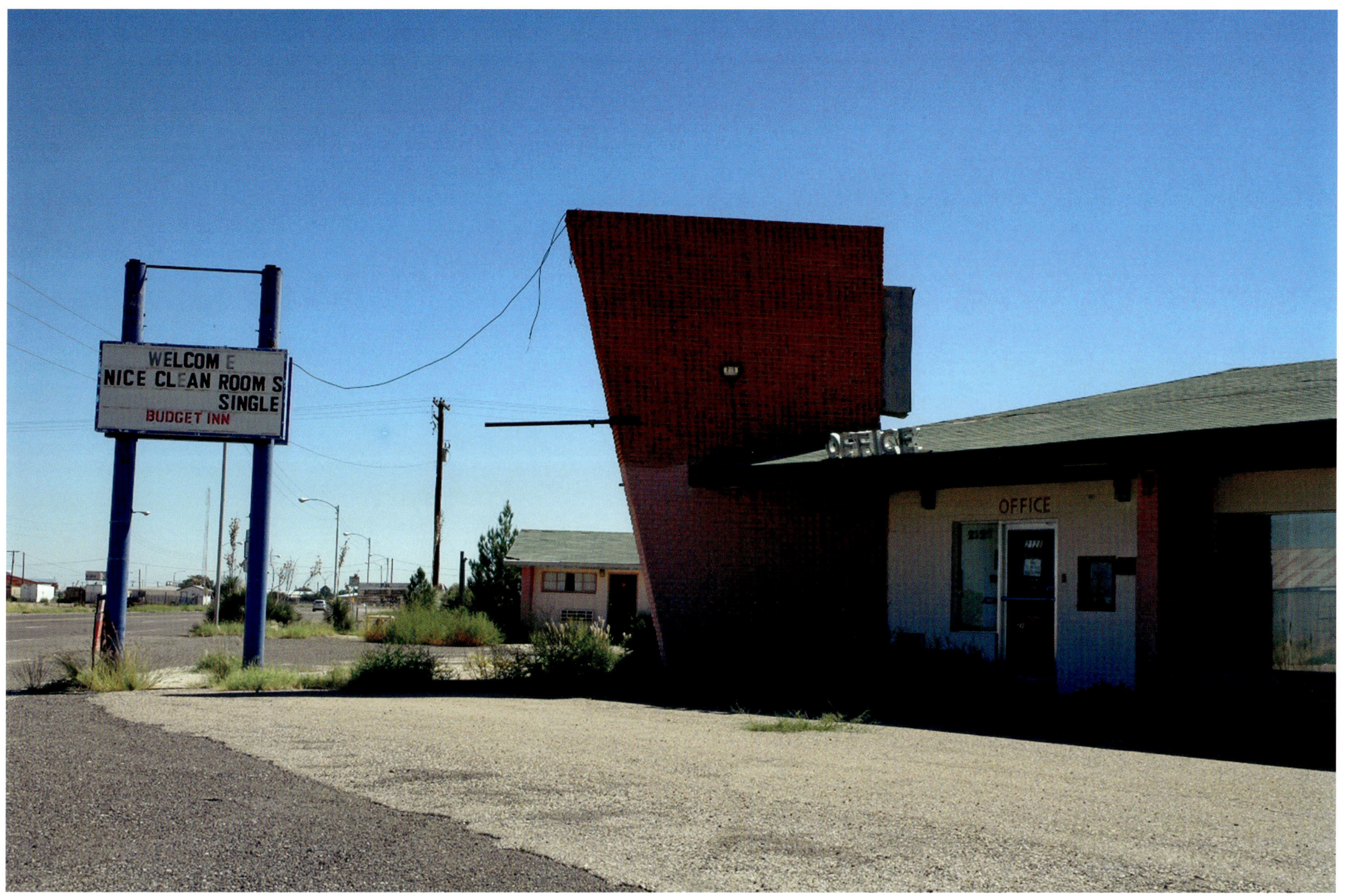
WELCOM E
NICE CLEAN ROOM S
SINGLE
BUDGET INN
OFFICE
OFFICE

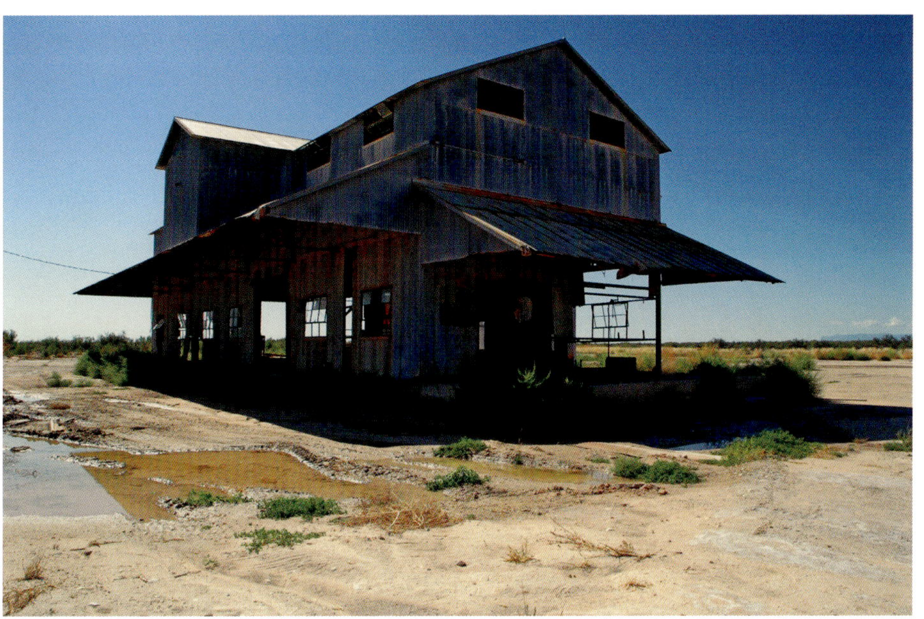

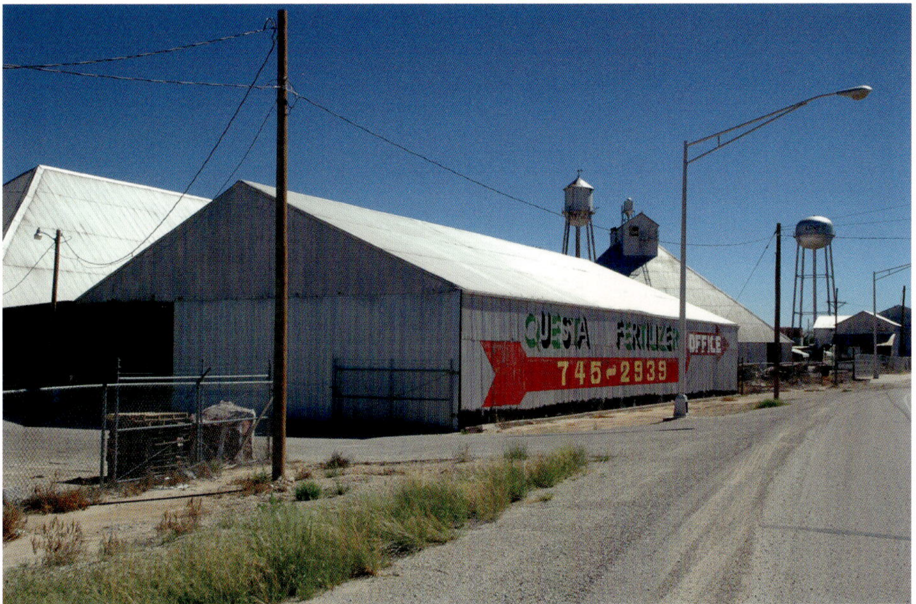
QUESTA  FERTILIZER  OFFICE
745-2939

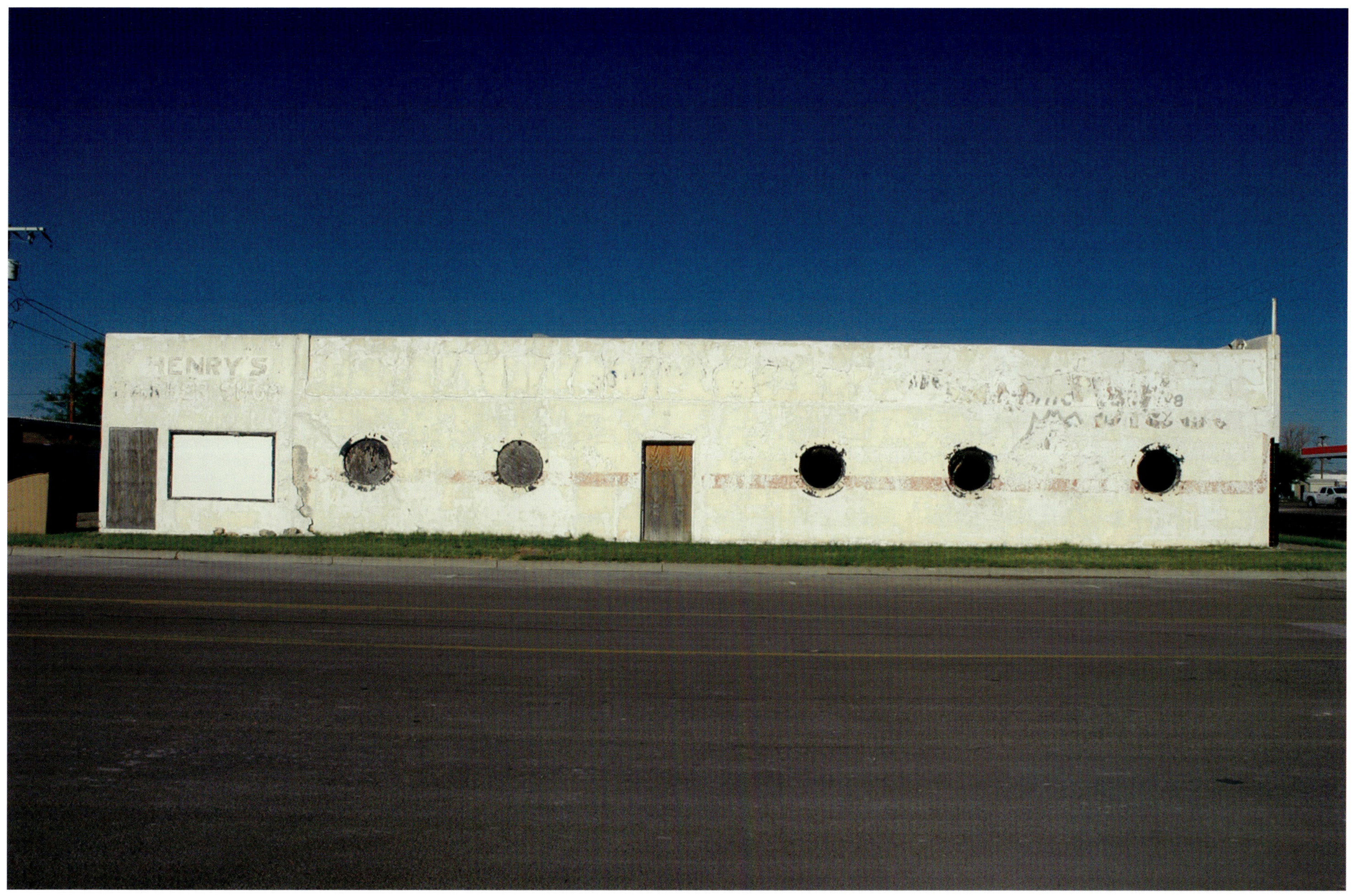
HENRY'S
BARBER SHOP

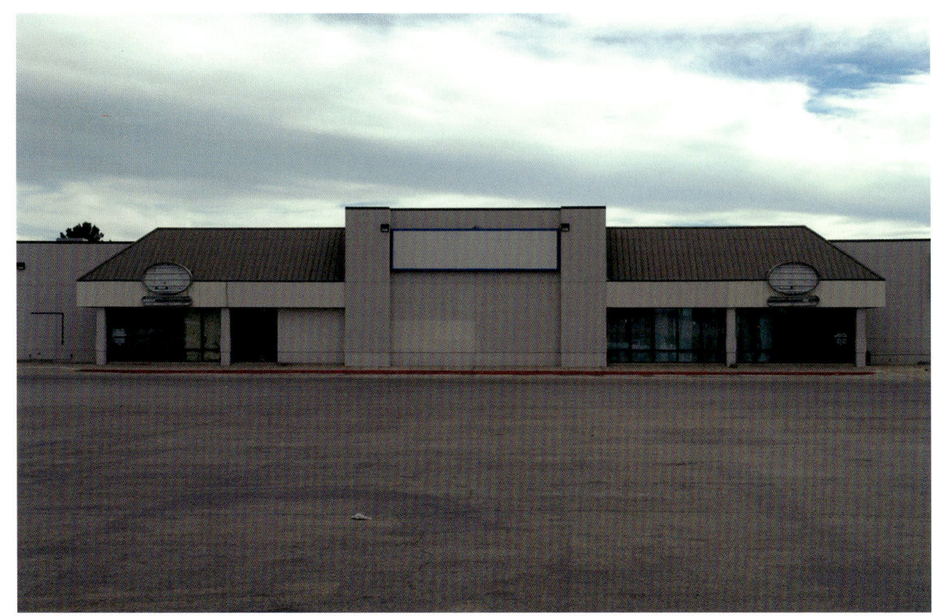
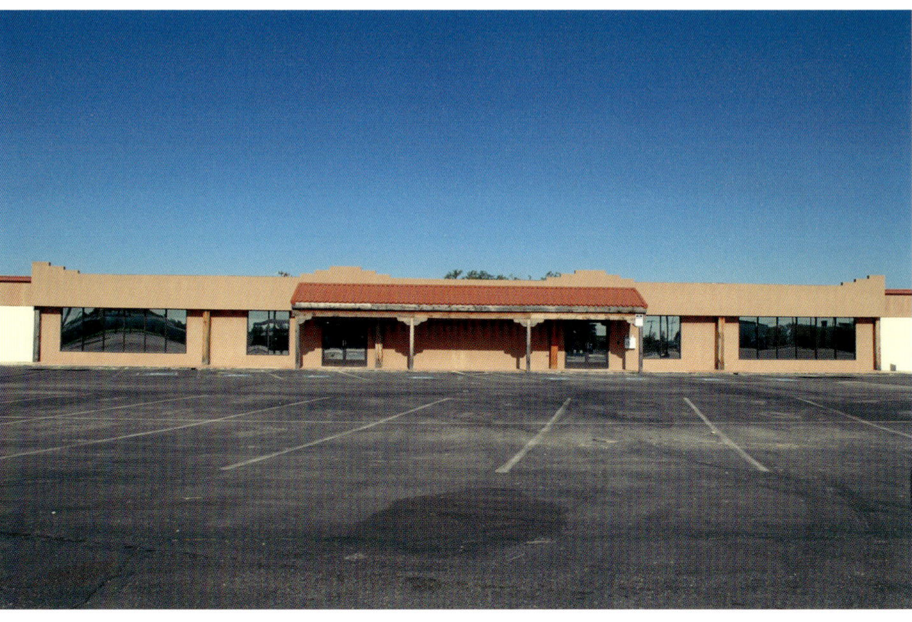
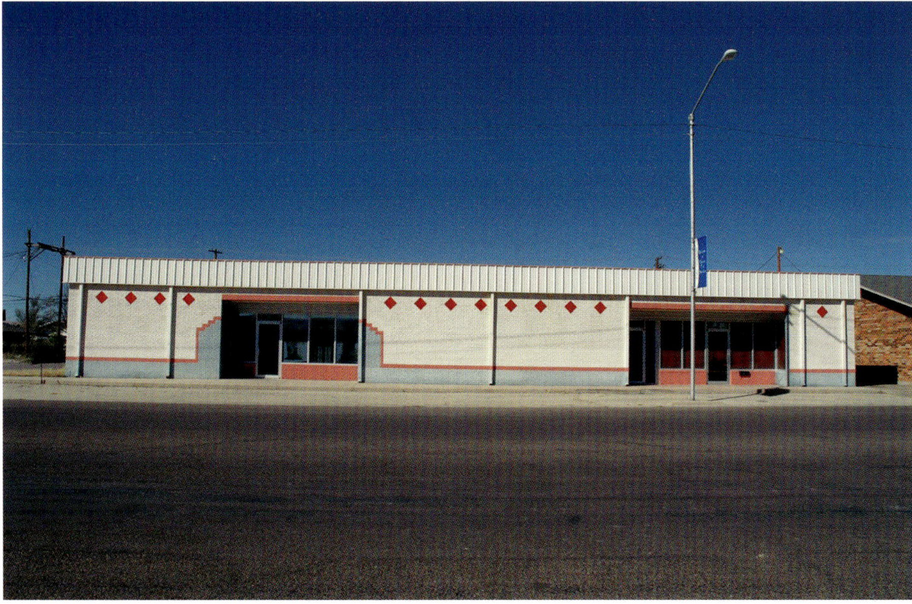
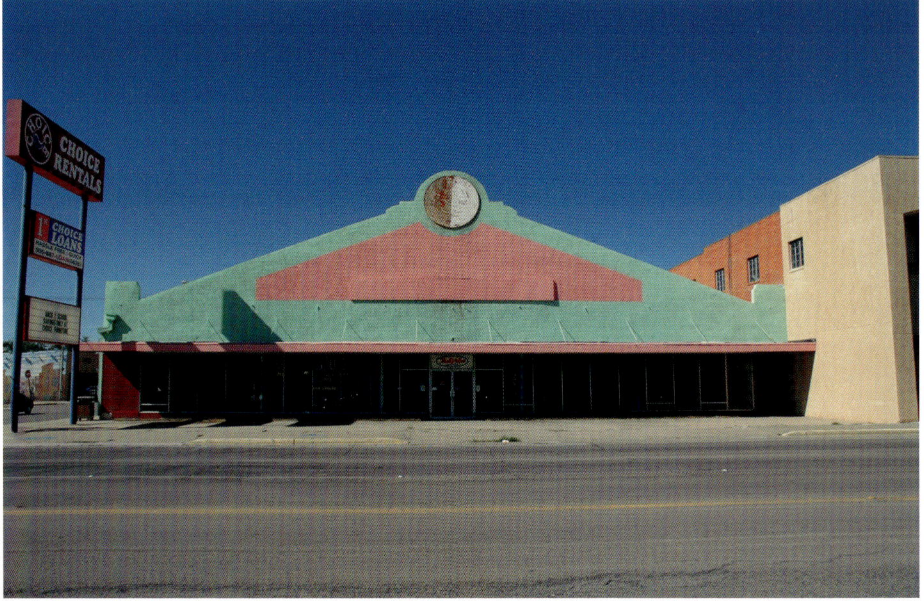

CHOICE
RENTALS
CHOICE
LOANS

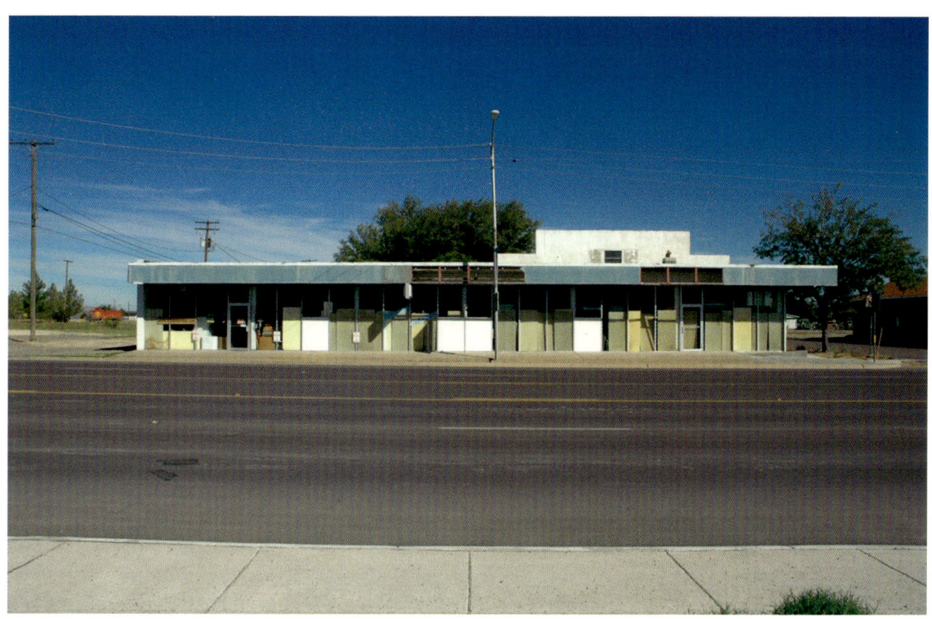

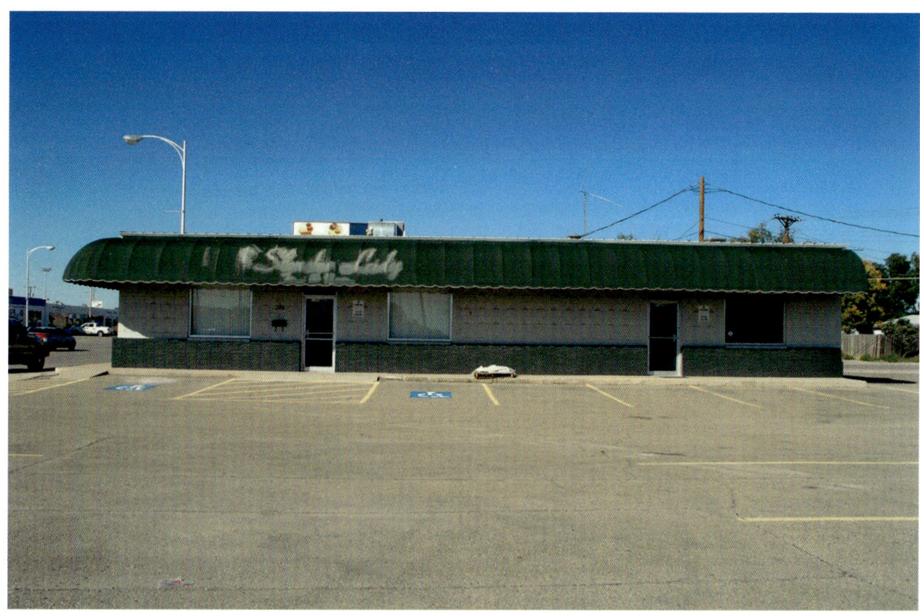

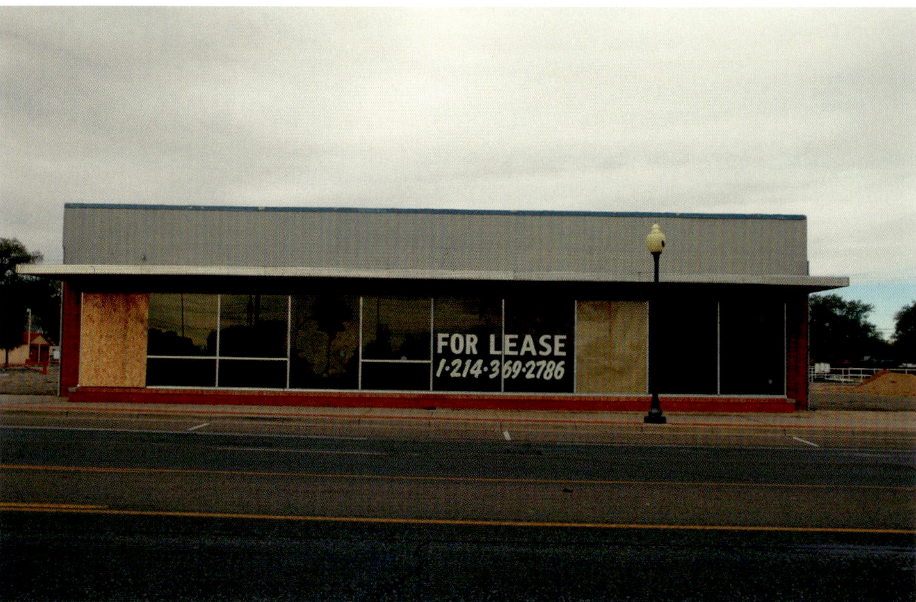
FOR LEASE
1·214·369·2786

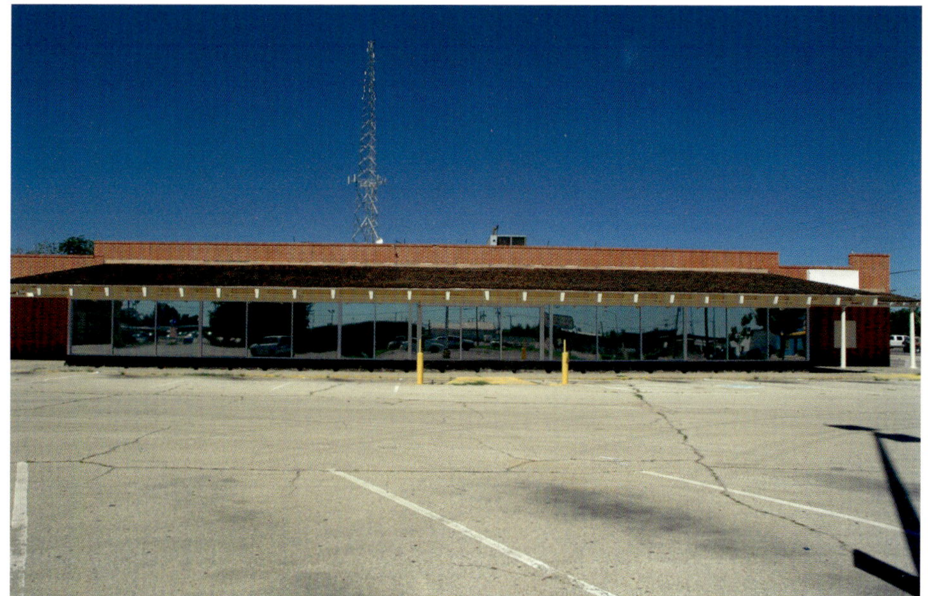

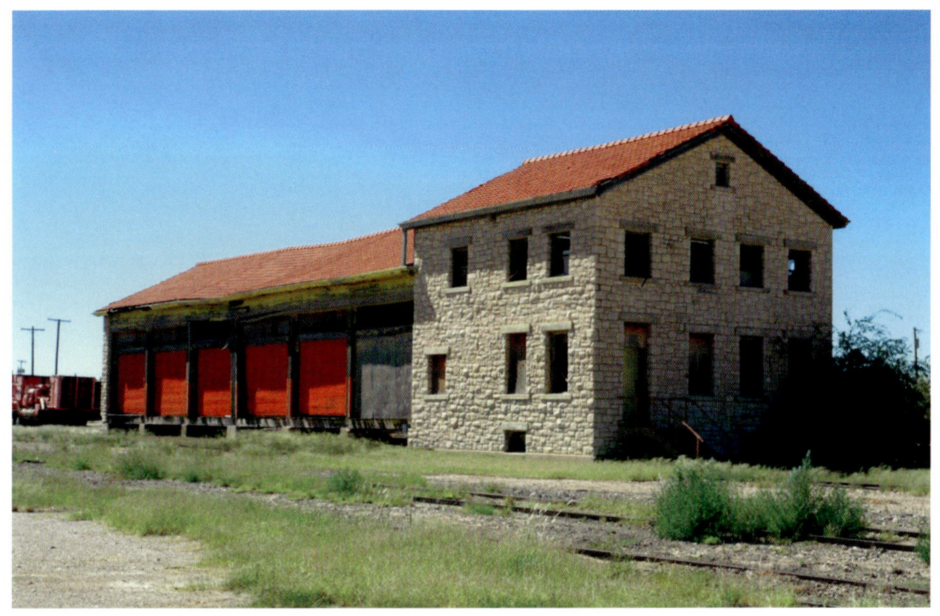

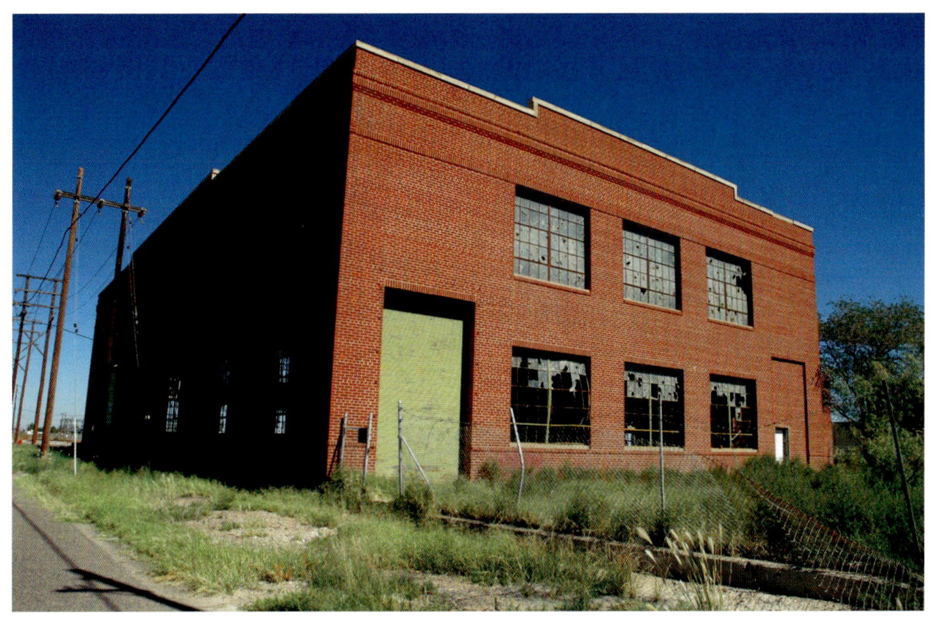

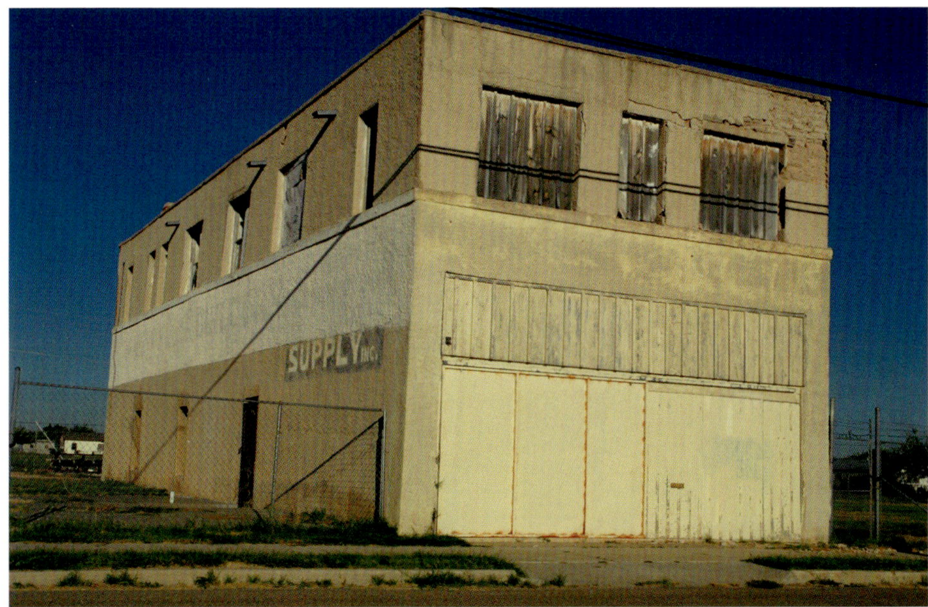

SUPPLY INC.

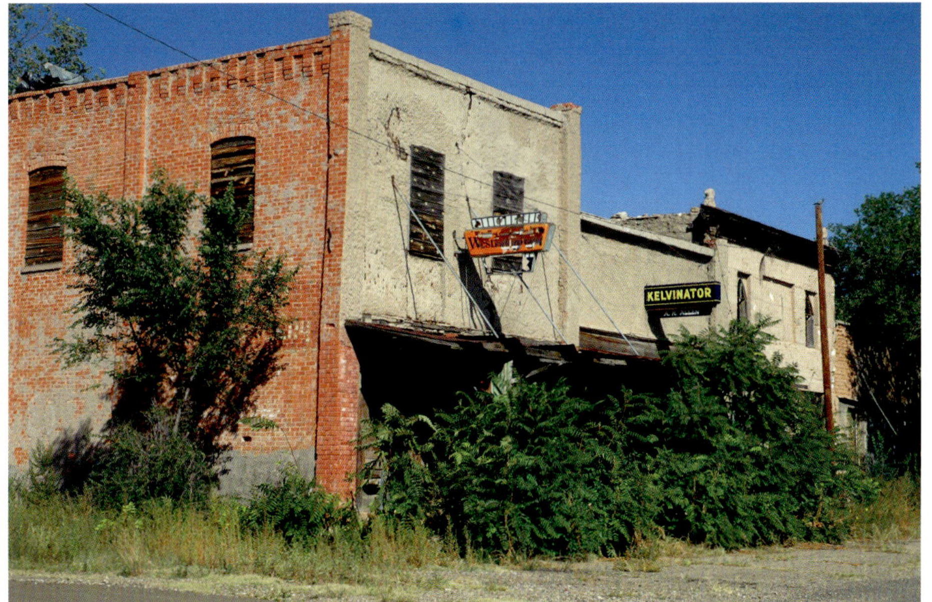

KELVINATOR

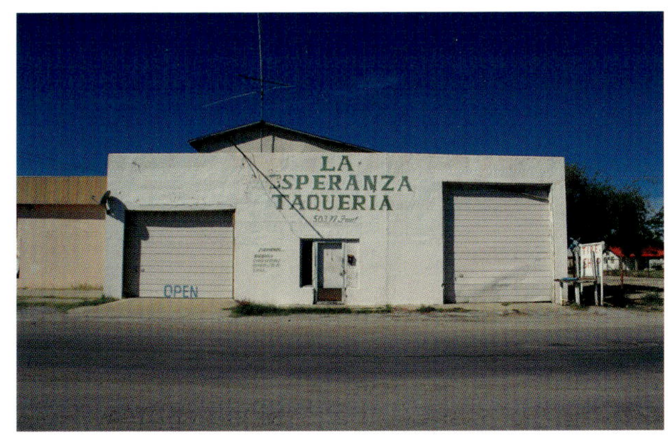

LA
ESPERANZA
TAQUERIA
OPEN


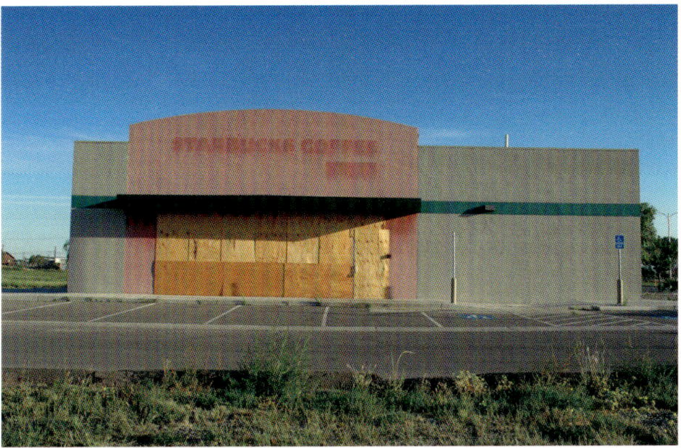

STARBUCKS COFFEE


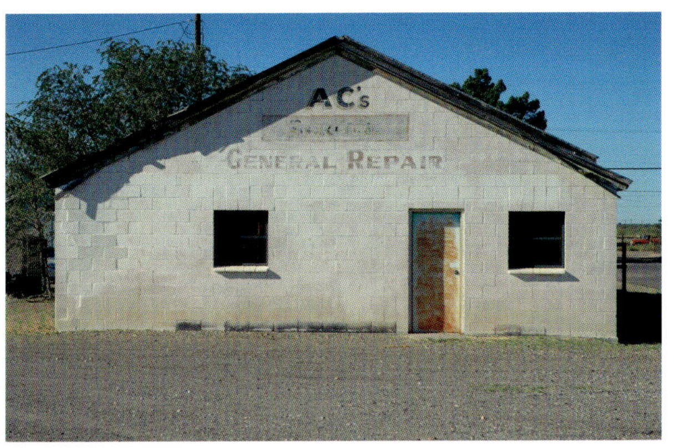

AC's
GENERAL REPAIR


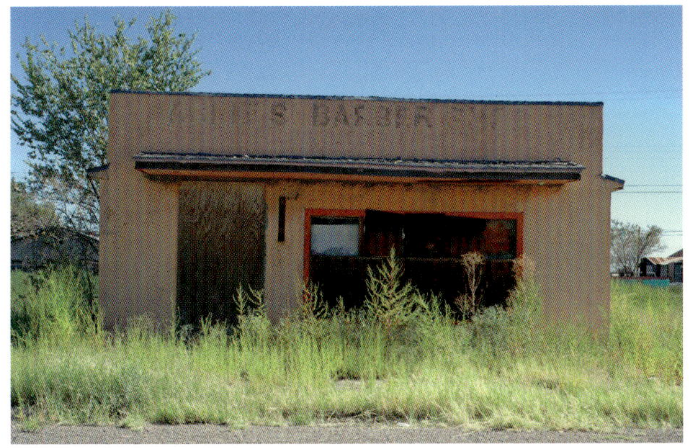

BARBER SH


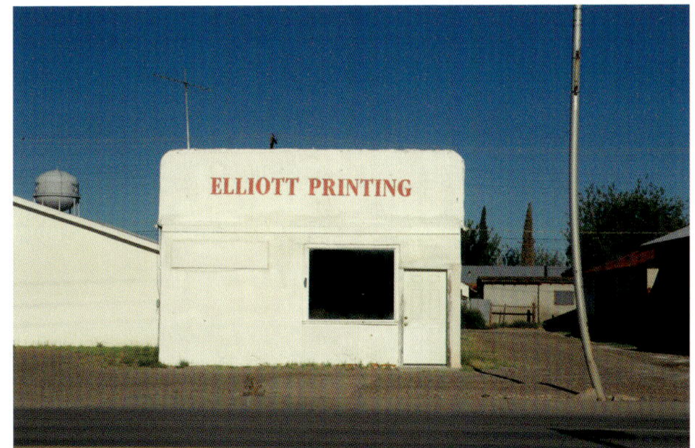

ELLIOTT PRINTING


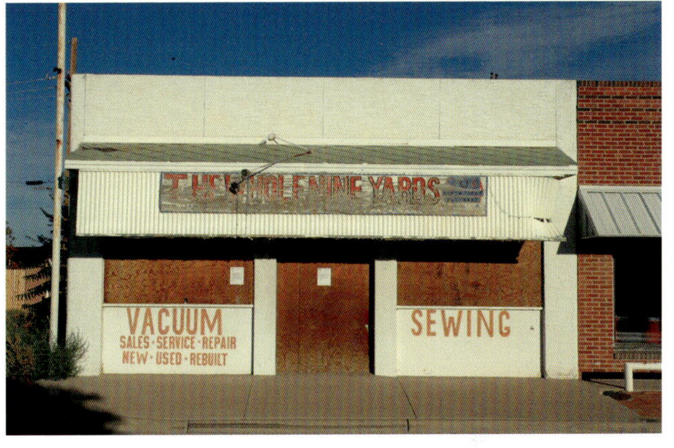

VACUUM
SALES · SERVICE · REPAIR
NEW · USED · REBUILT
SEWING


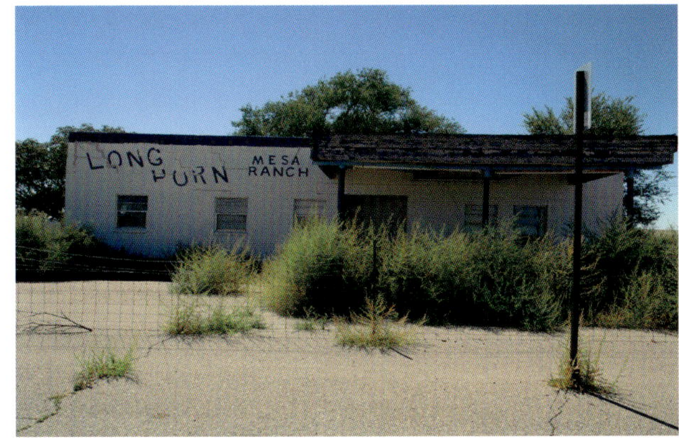

LONG
HORN
MESA
RANCH


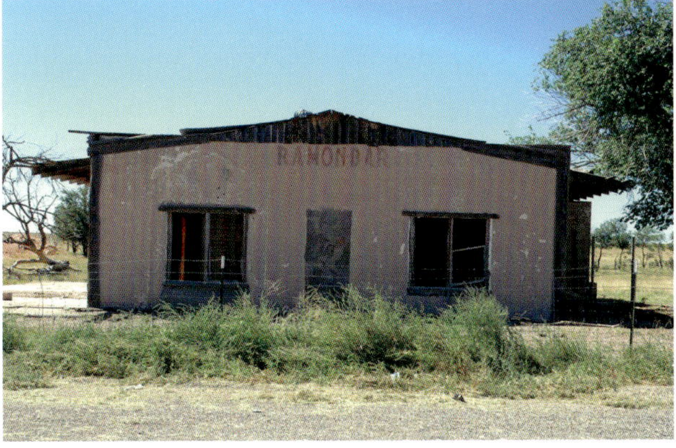

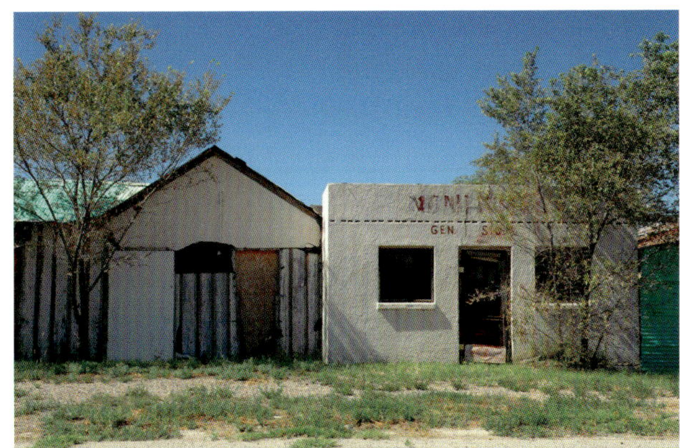

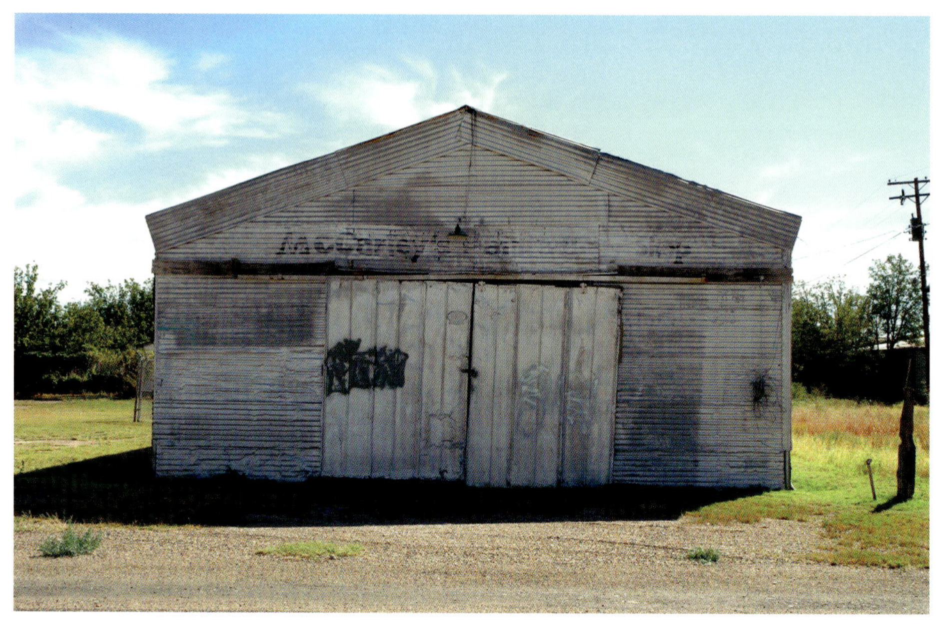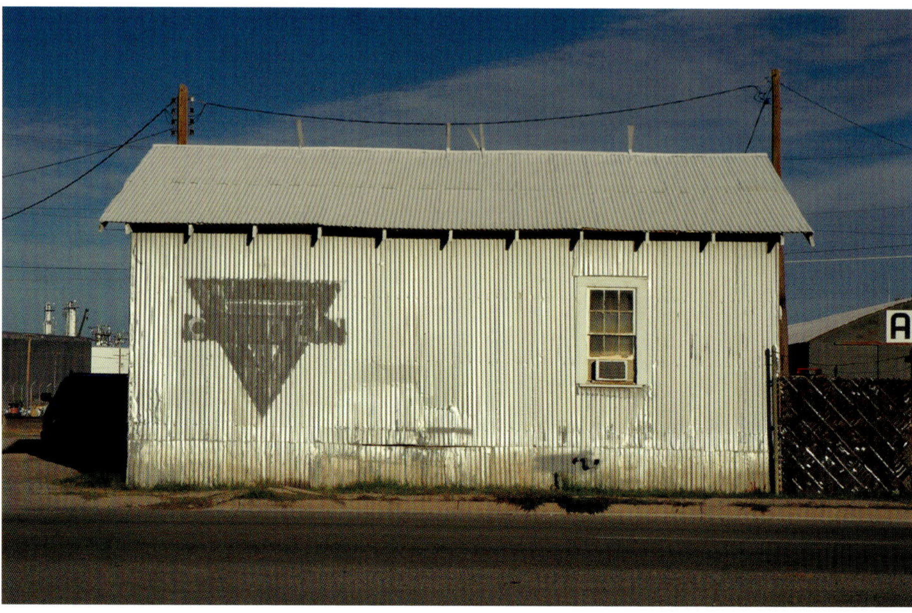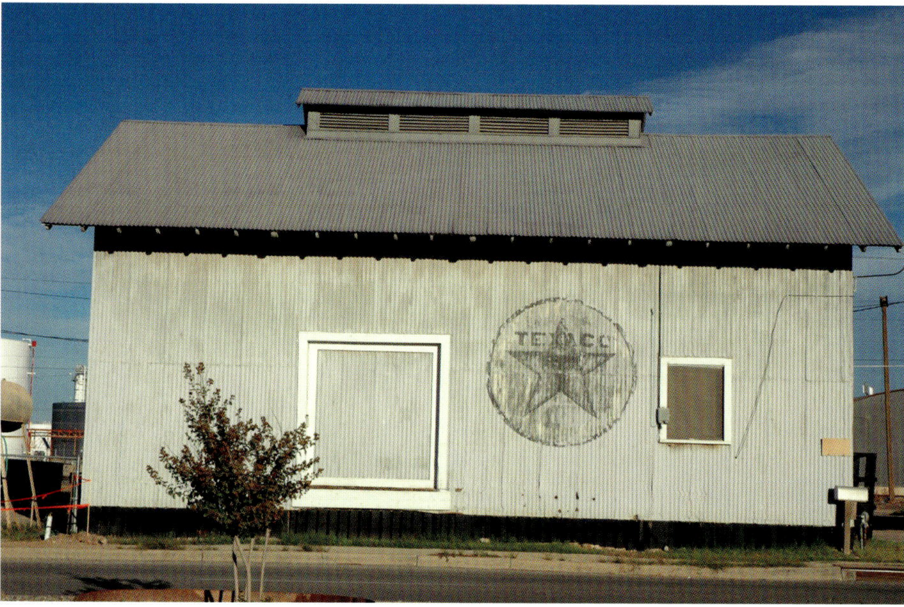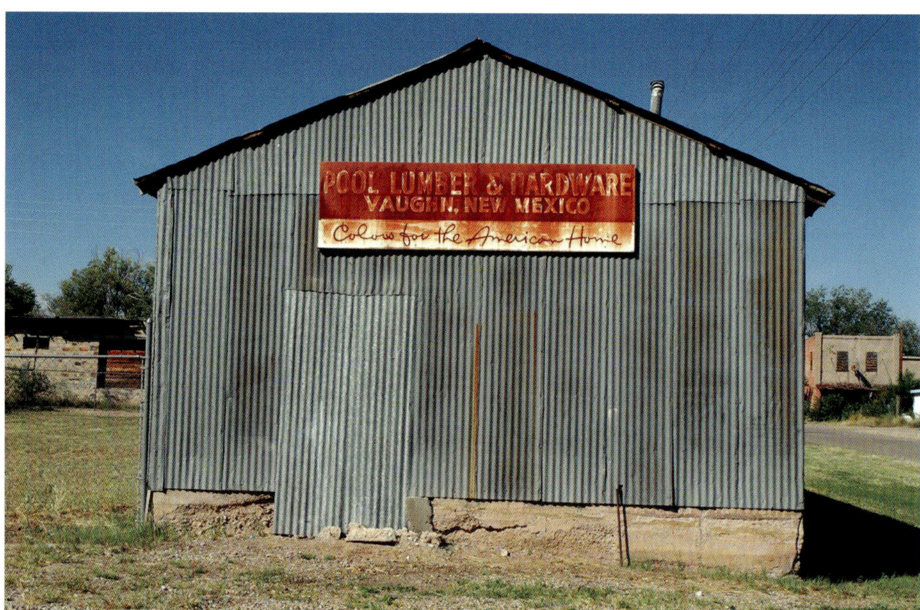
POOL LUMBER & HARDWARE
VAUGHN, NEW MEXICO
Colors for the American Home

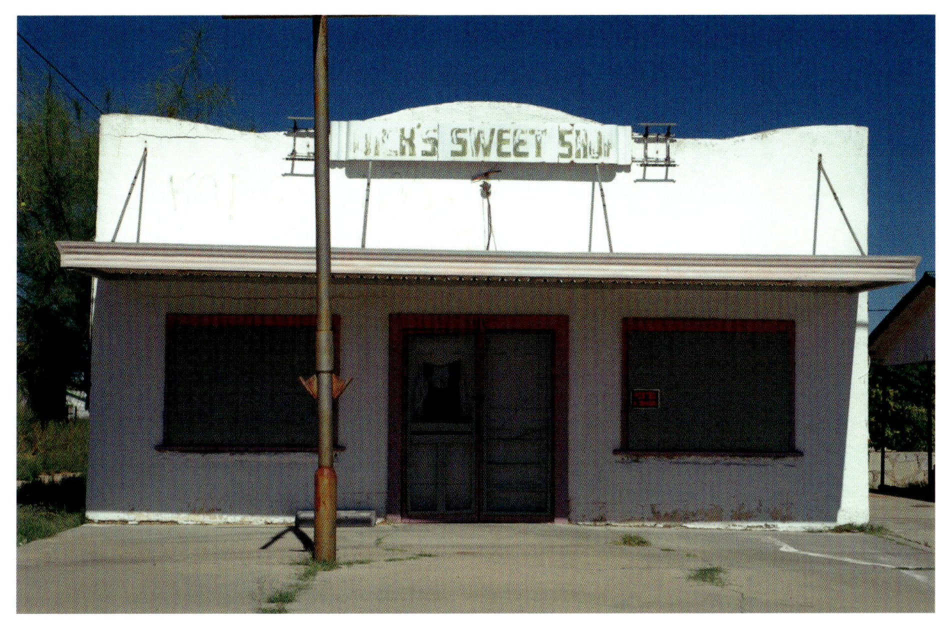

DICK'S SWEET SHOP


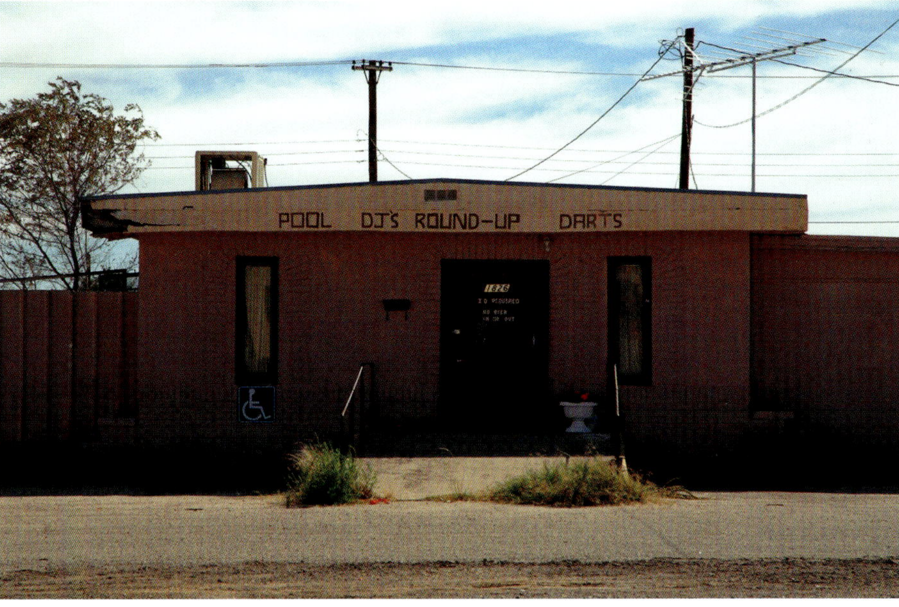

POOL   DJ'S ROUND-UP   DARTS


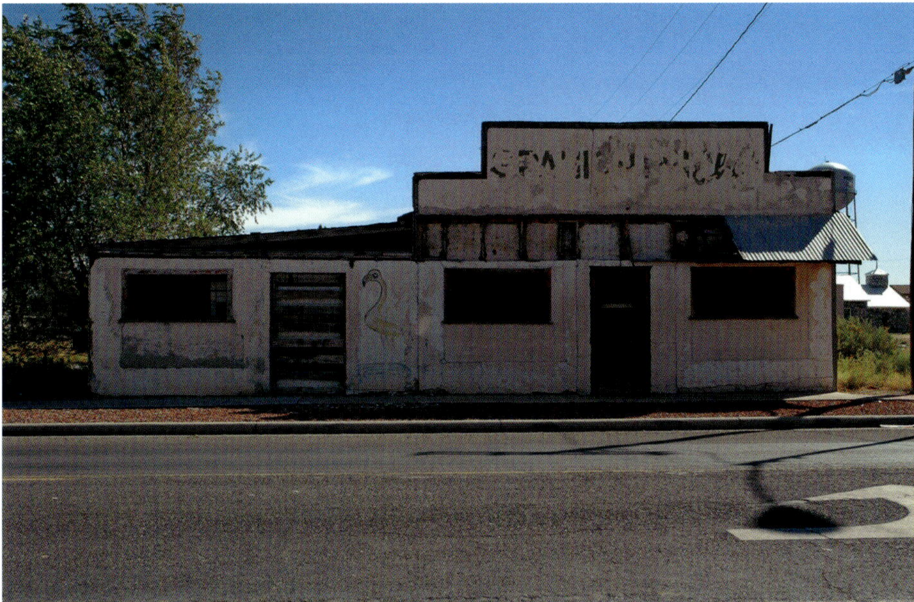

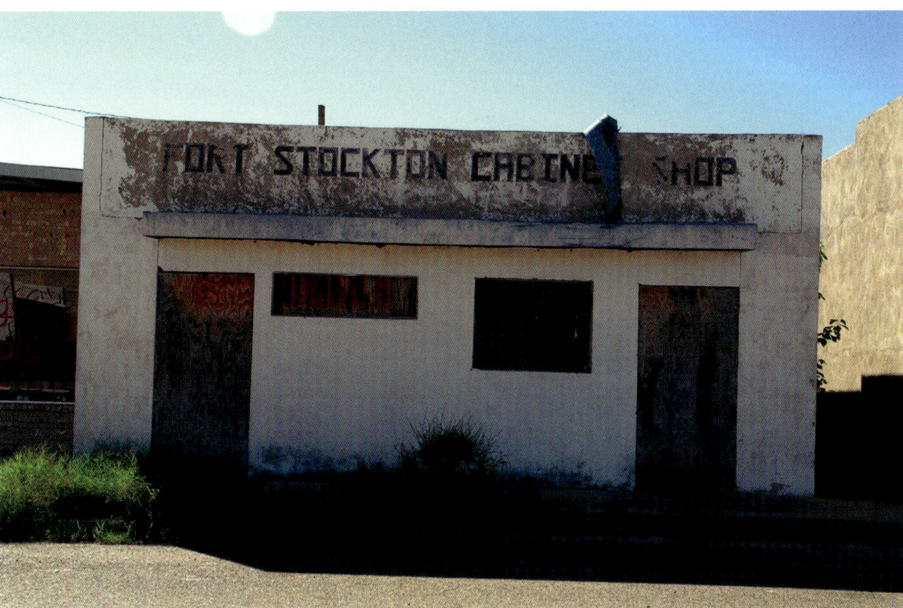

FORT STOCKTON CABINET SHOP

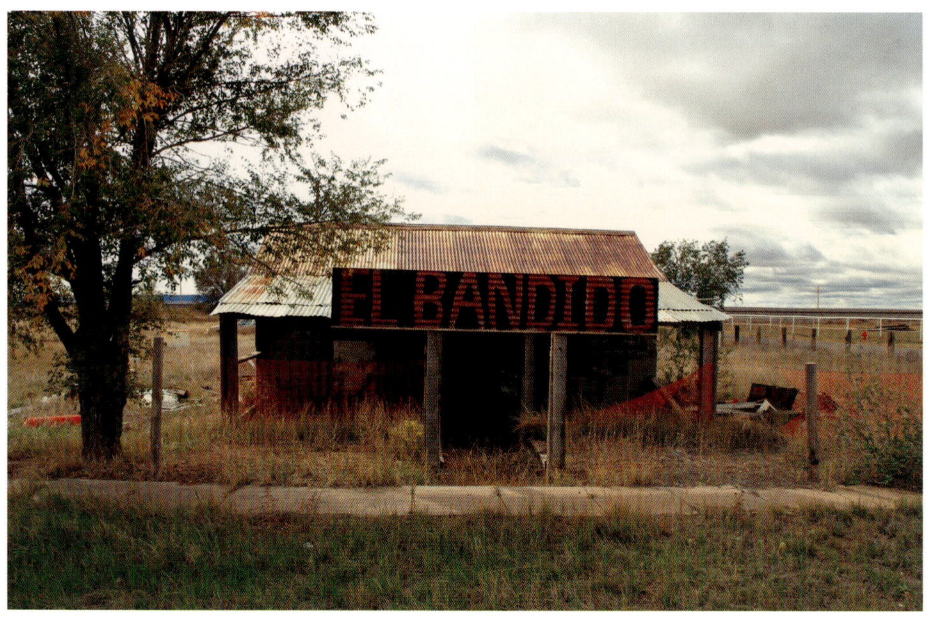
EL BANDIDO

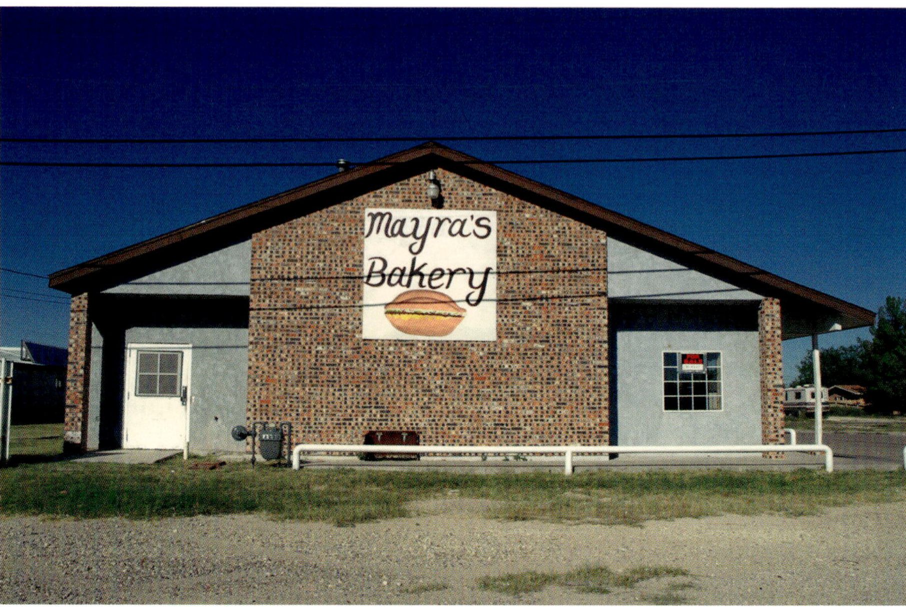
Mayra's
Bakery

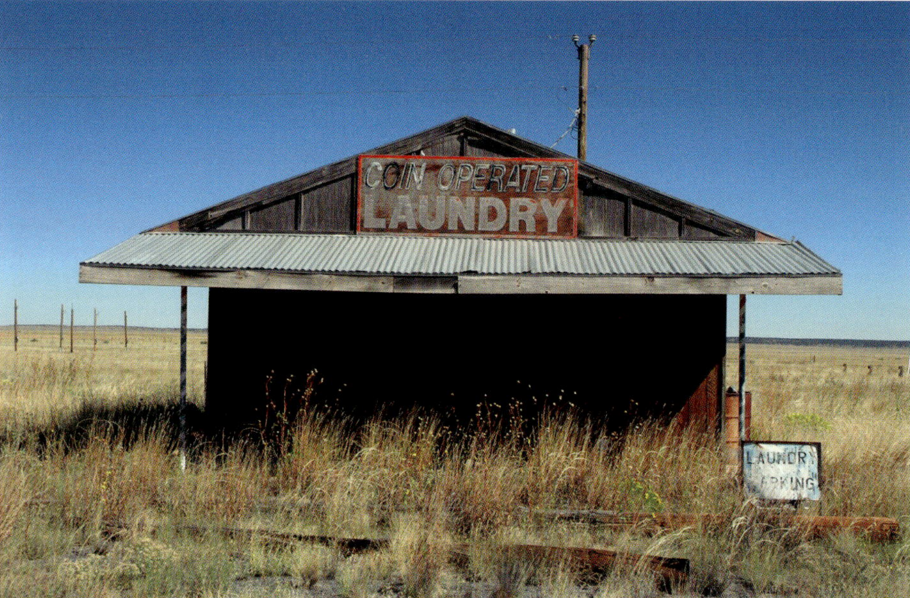
COIN OPERATED
LAUNDRY
LAUNDRY
PARKING

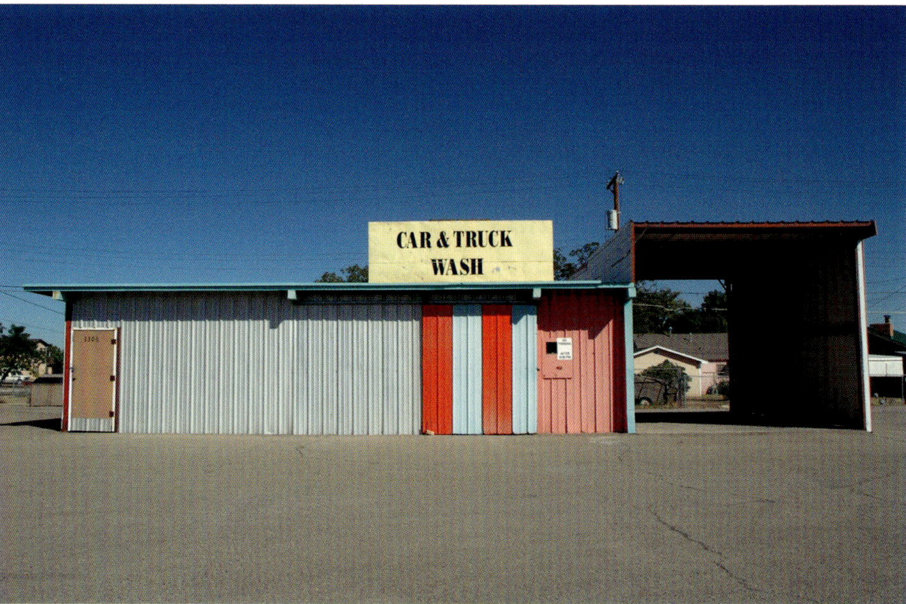
CAR & TRUCK
WASH

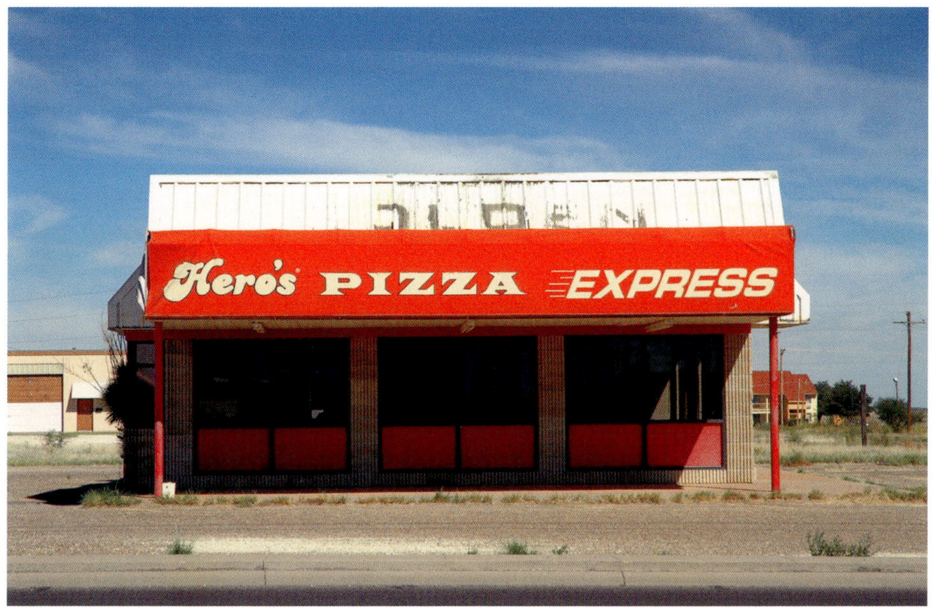

Hero's PIZZA EXPRESS


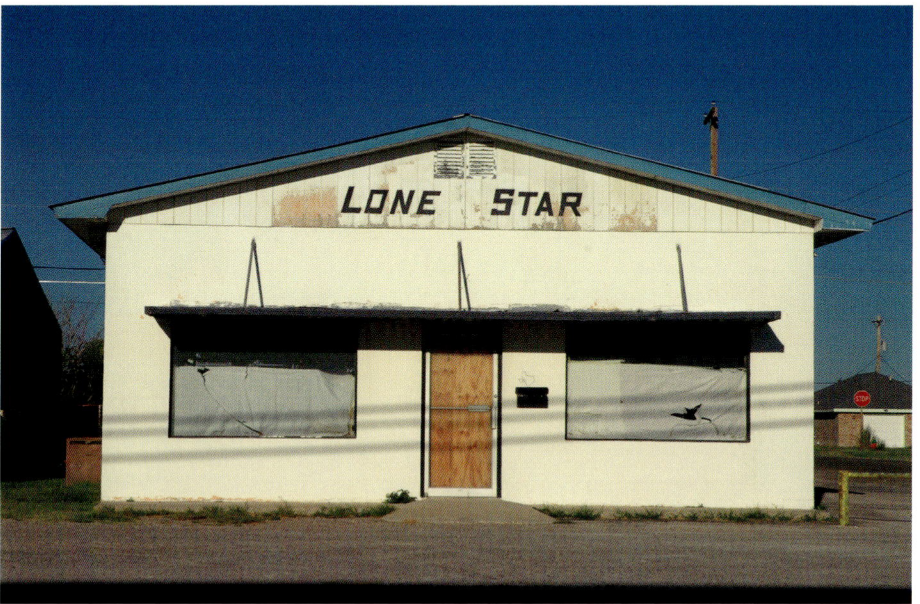

LONE STAR


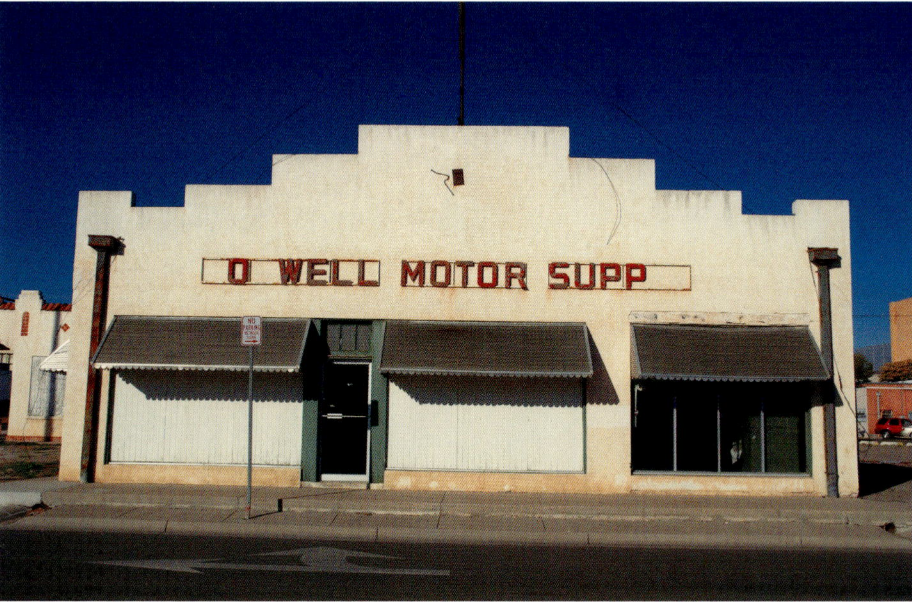

O WELL MOTOR SUPP


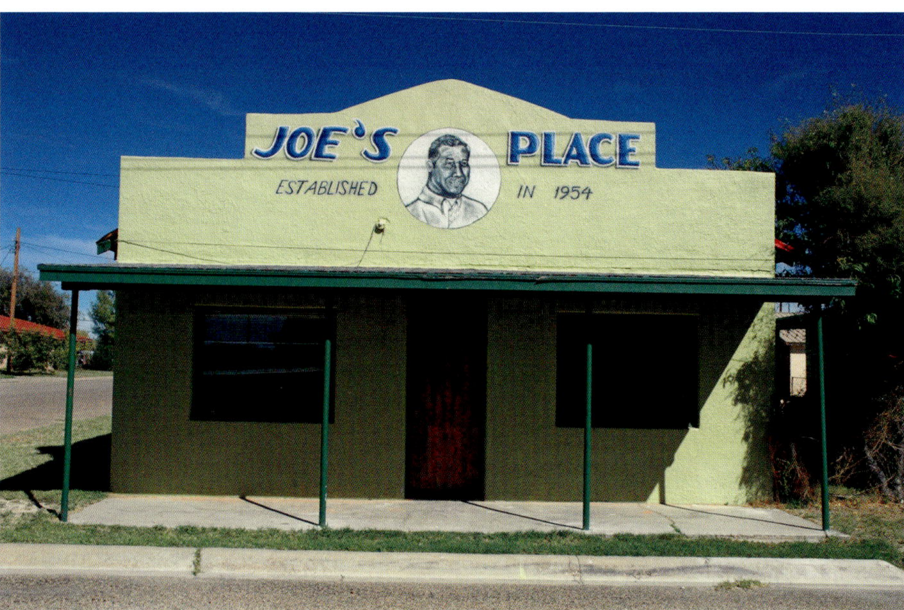

JOE'S PLACE
ESTABLISHED IN 1954

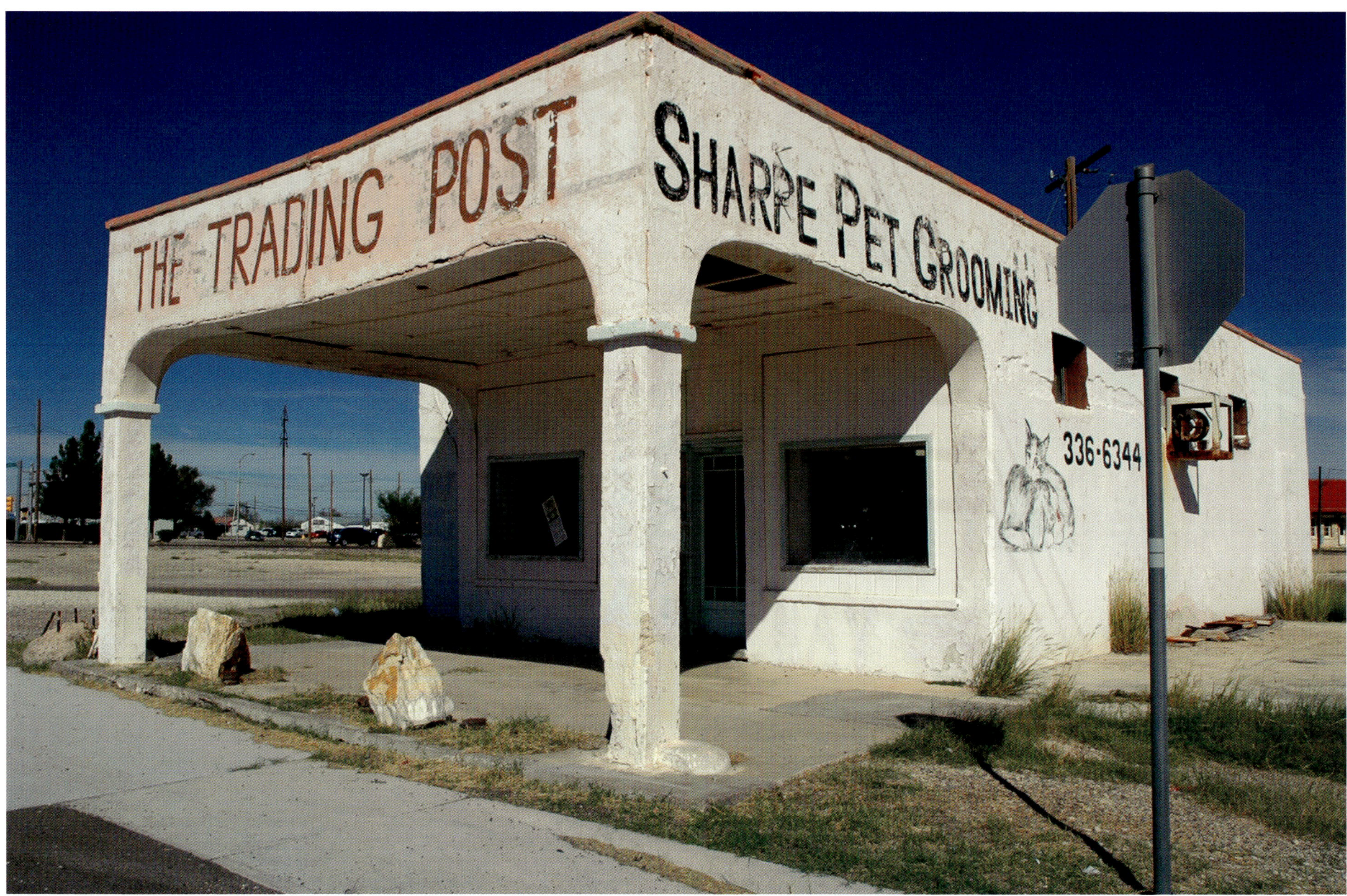
THE TRADING POST
SHARPE PET GROOMING
336-6344

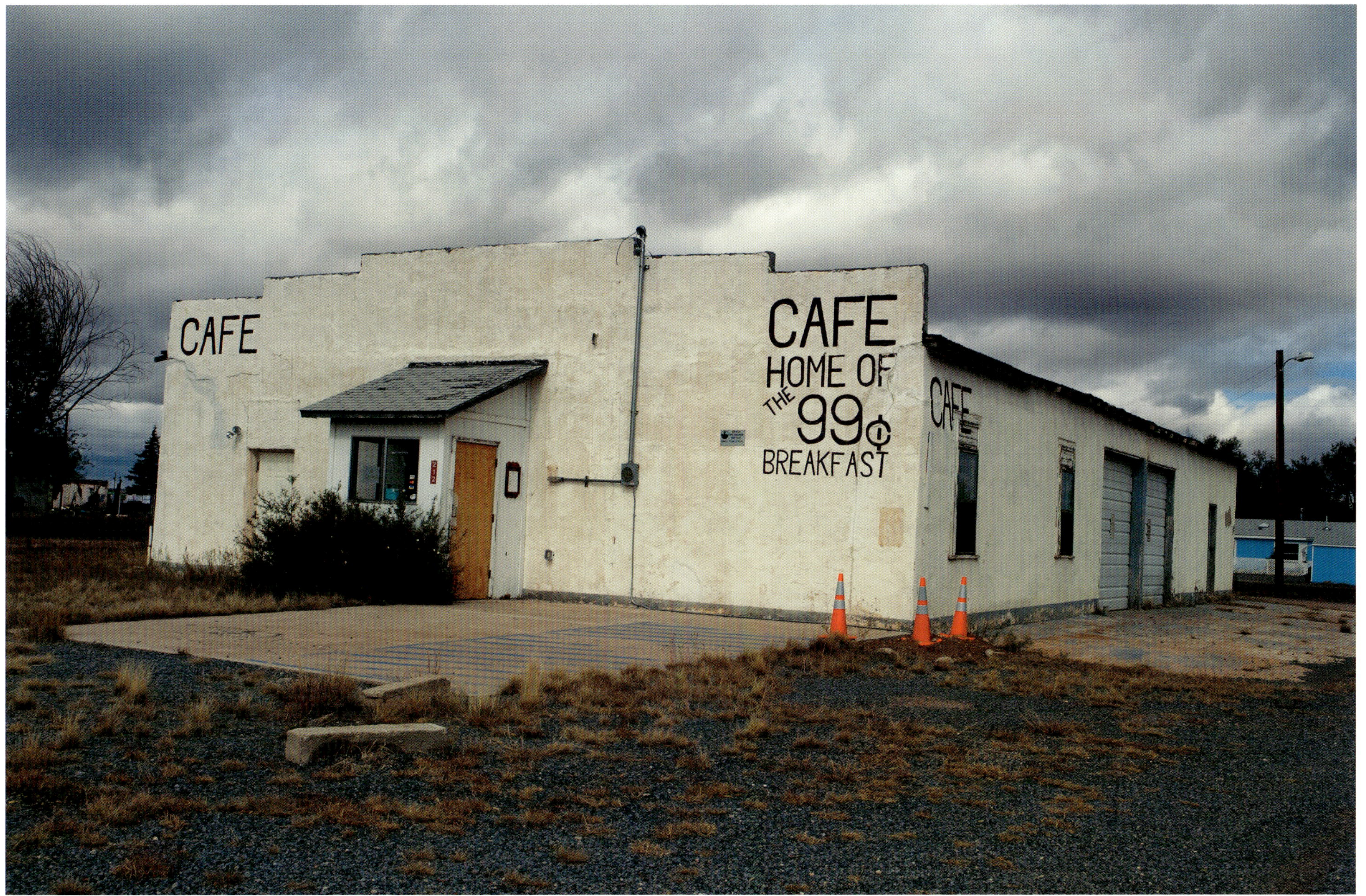
CAFE
CAFE
HOME OF
THE 99¢
BREAKFAST
CAFE

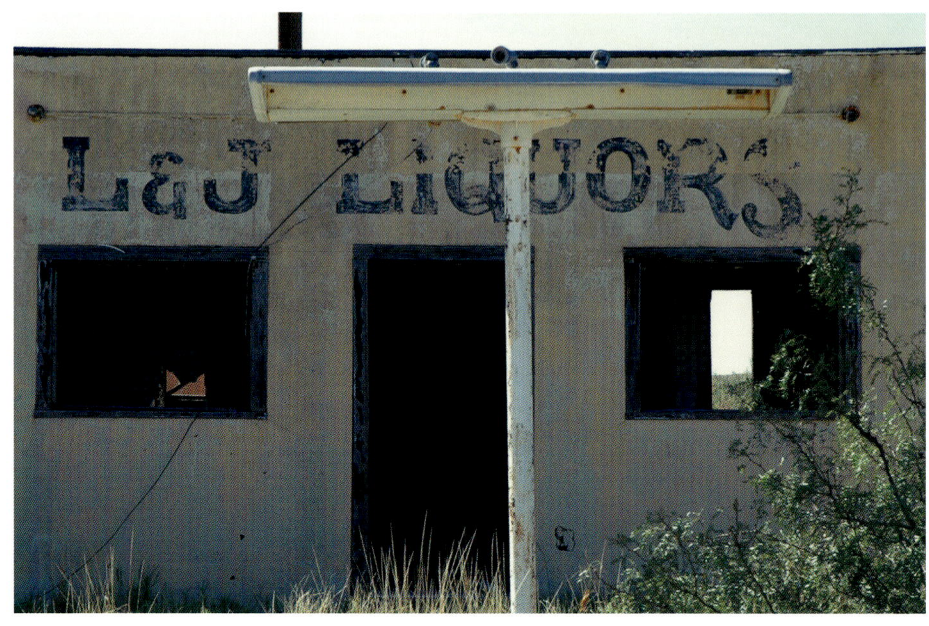

L&J LIQUORS


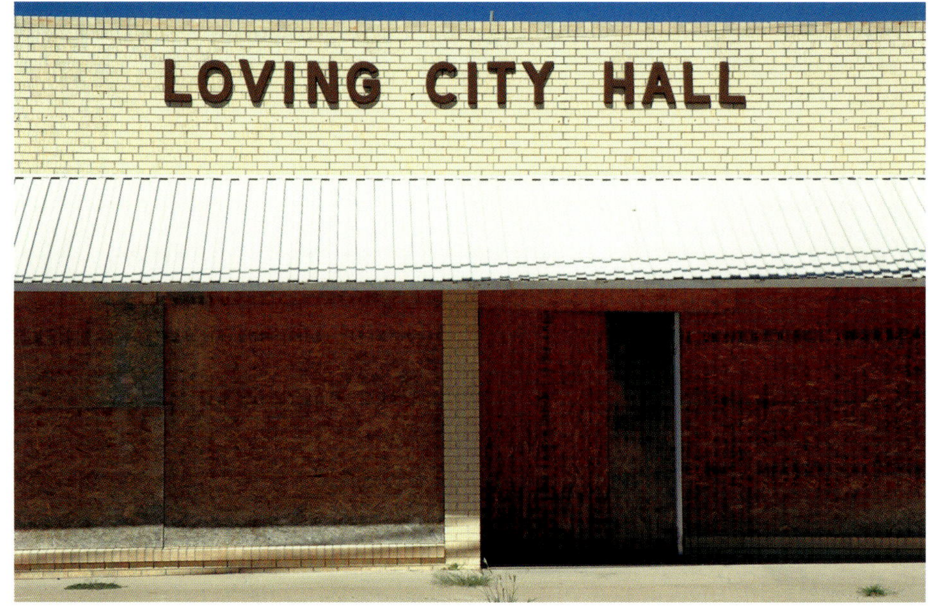

LOVING CITY HALL

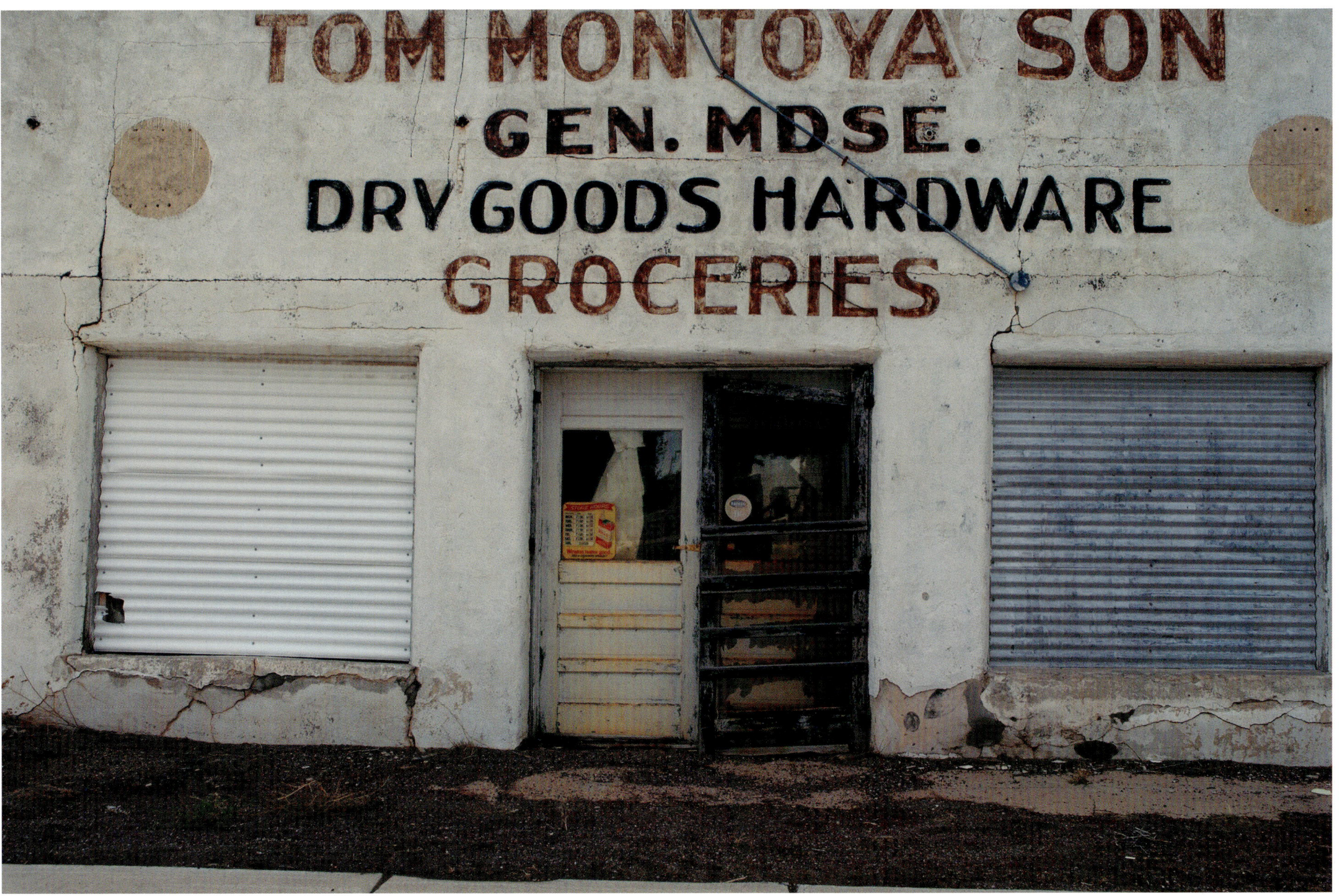

TOM MONTOYA SON
GEN. MDSE.
DRY GOODS HARDWARE
GROCERIES

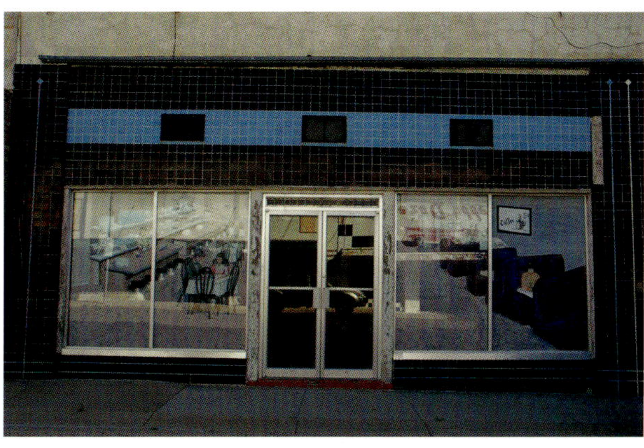
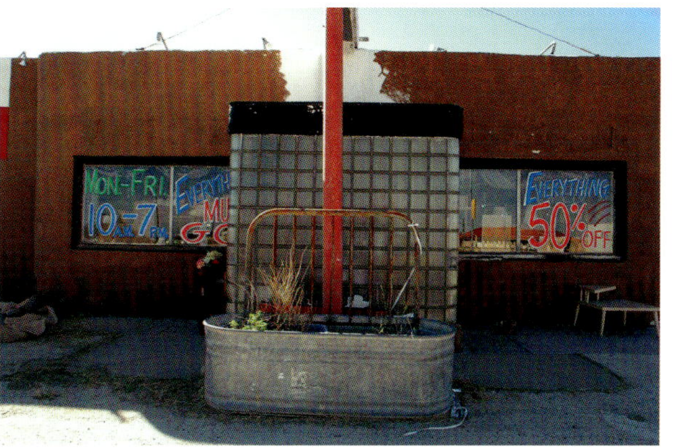

MON-FRI
10-7
EVERYTHING
50% OFF

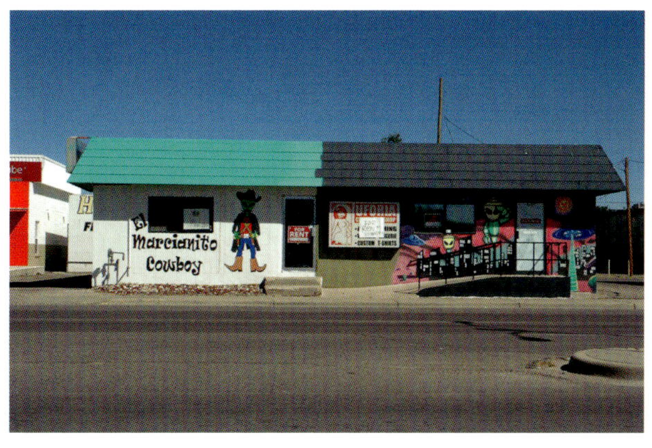

El
Marcianito
Cowboy

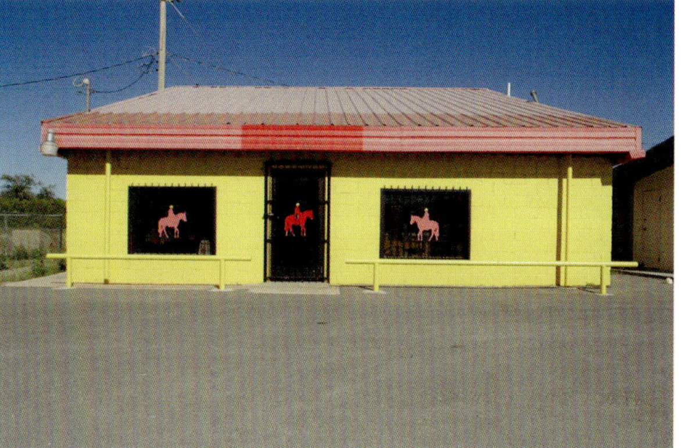
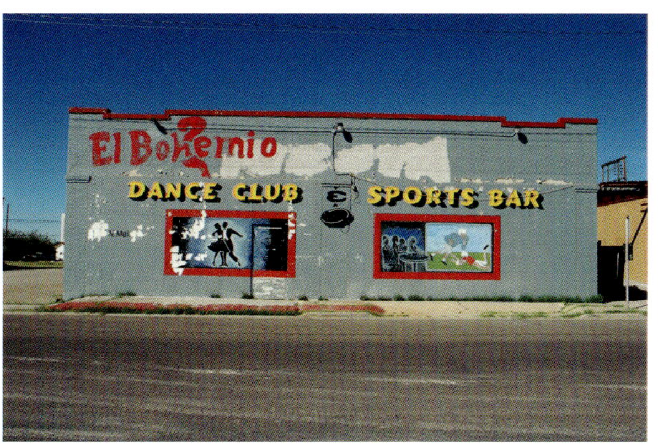

El Bohemio
DANCE CLUB
SPORTS BAR

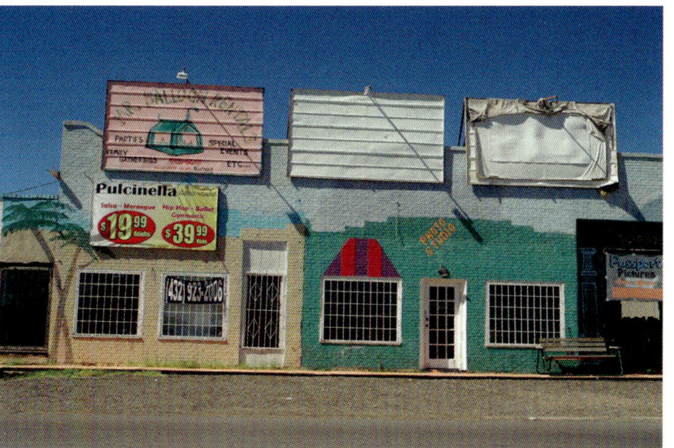

Pulcinella
$79.99
$39.99

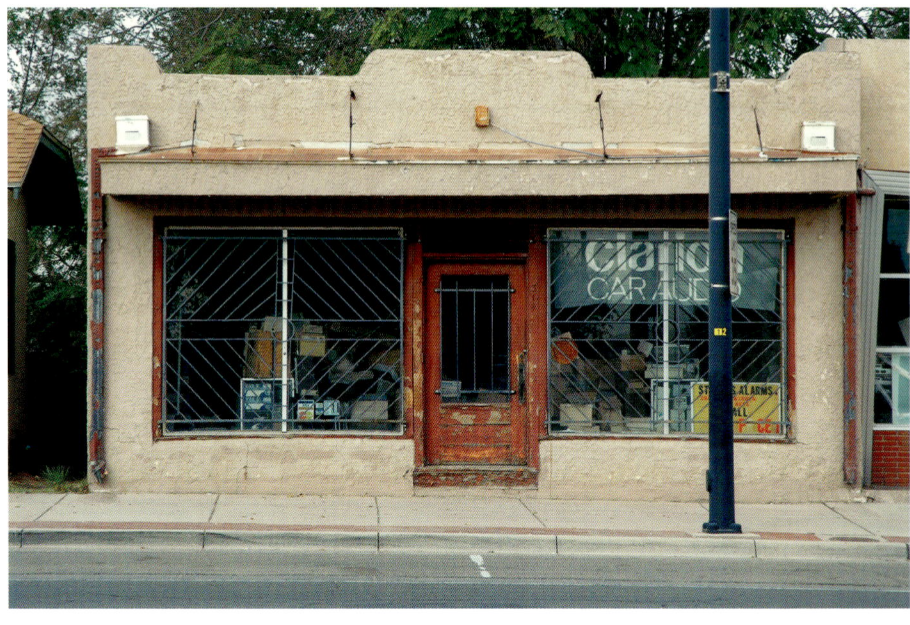

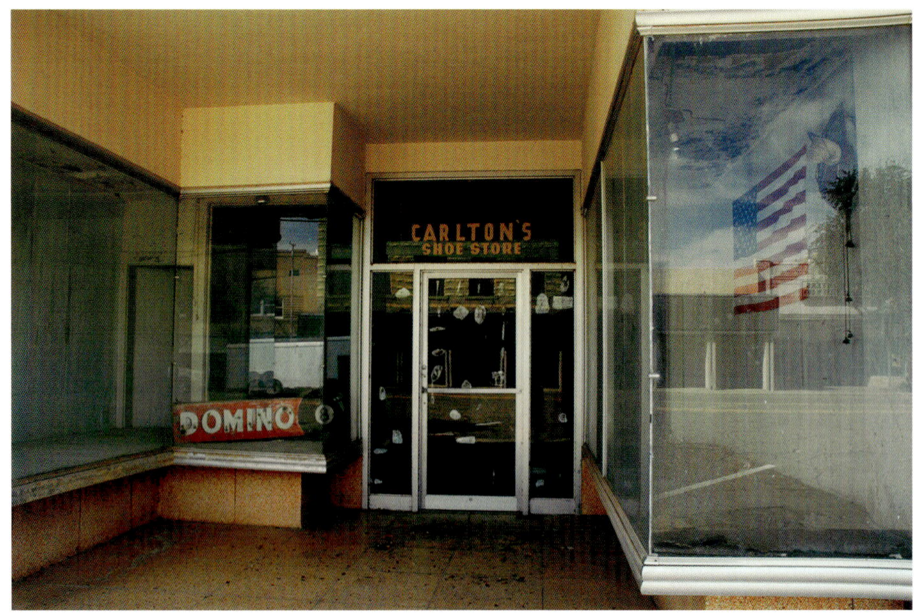
CARLTON'S
SHOE STORE
DOMINO

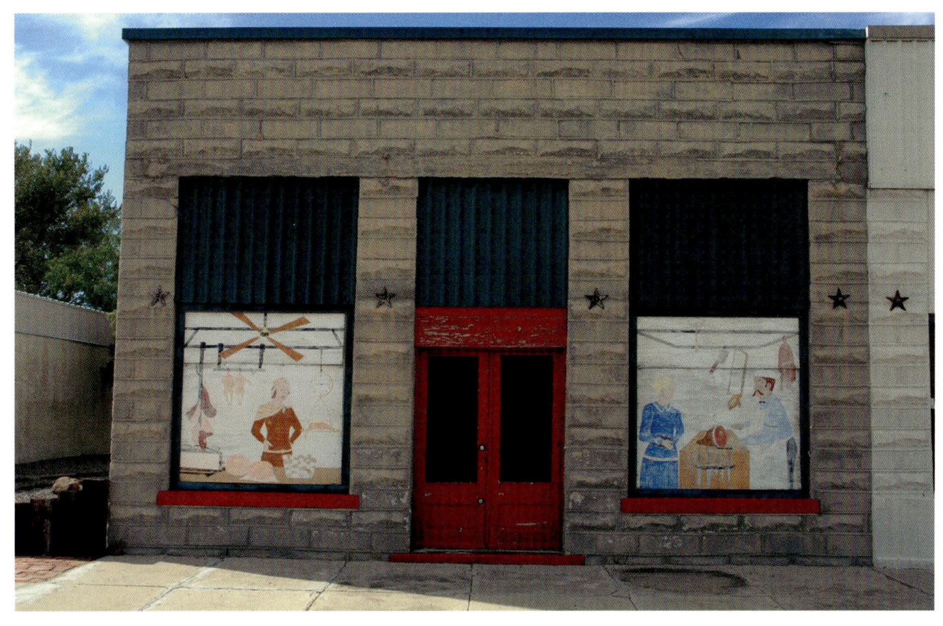
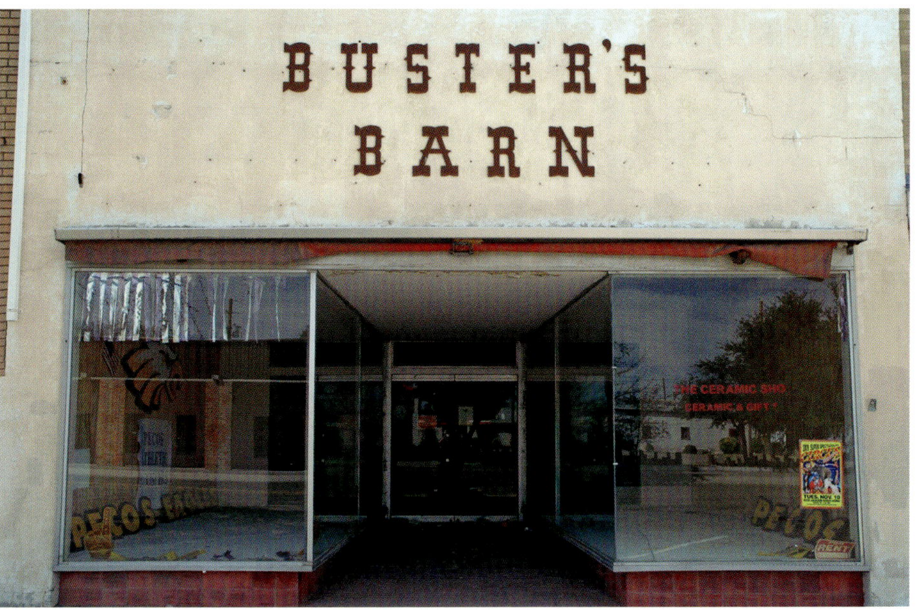

BUSTER'S
BARN

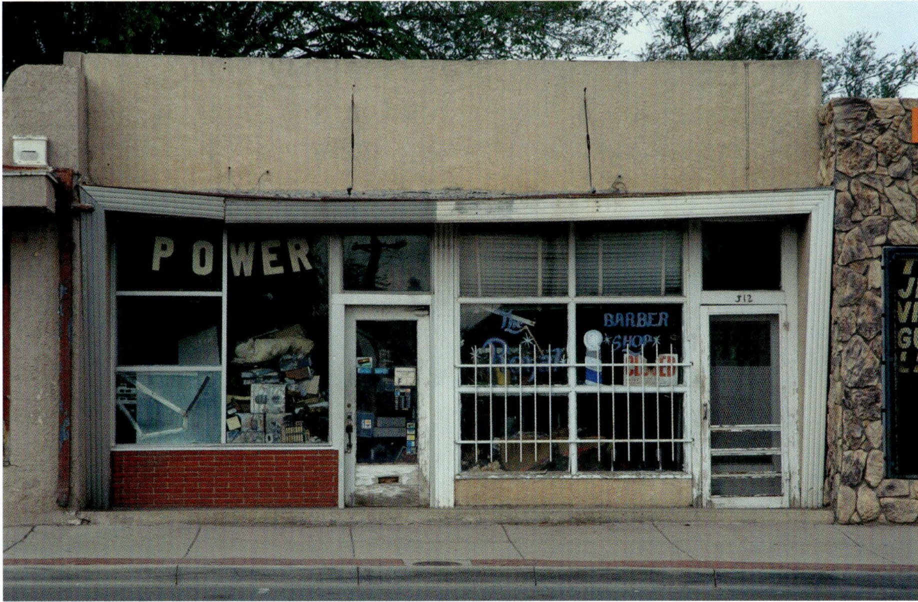

POWER

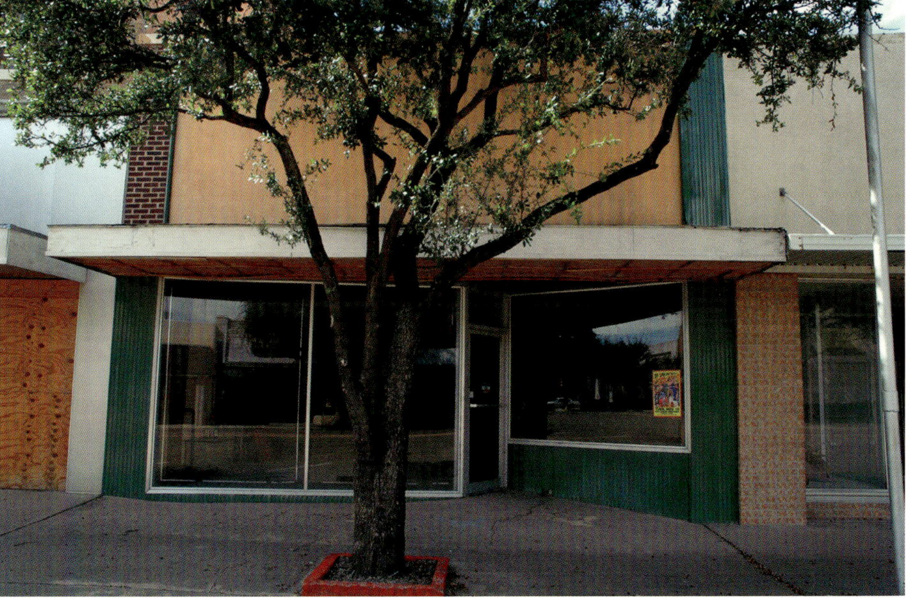

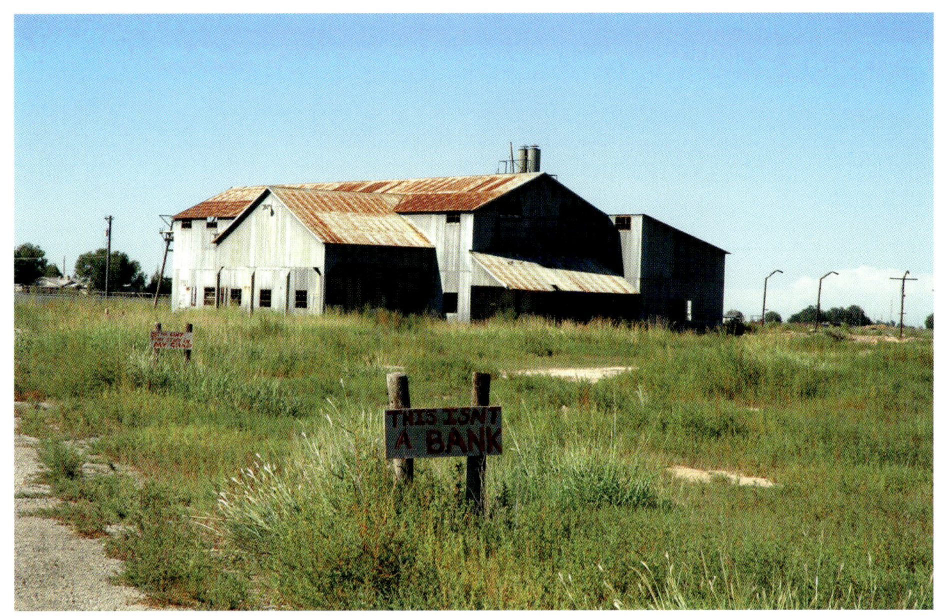
THIS ISN'T
A BANK

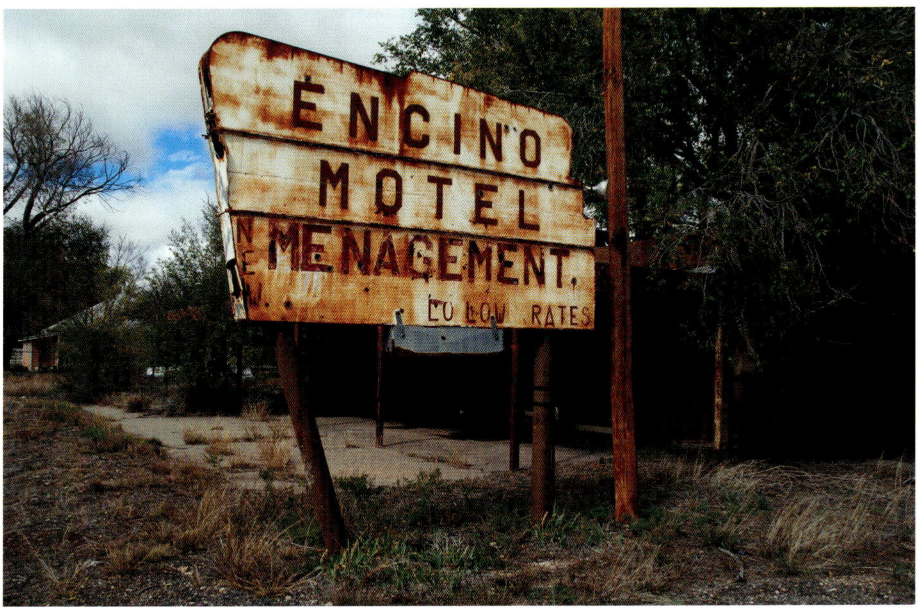
ENCINO
MOTEL
MENAGEMENT
LO LOW RATES

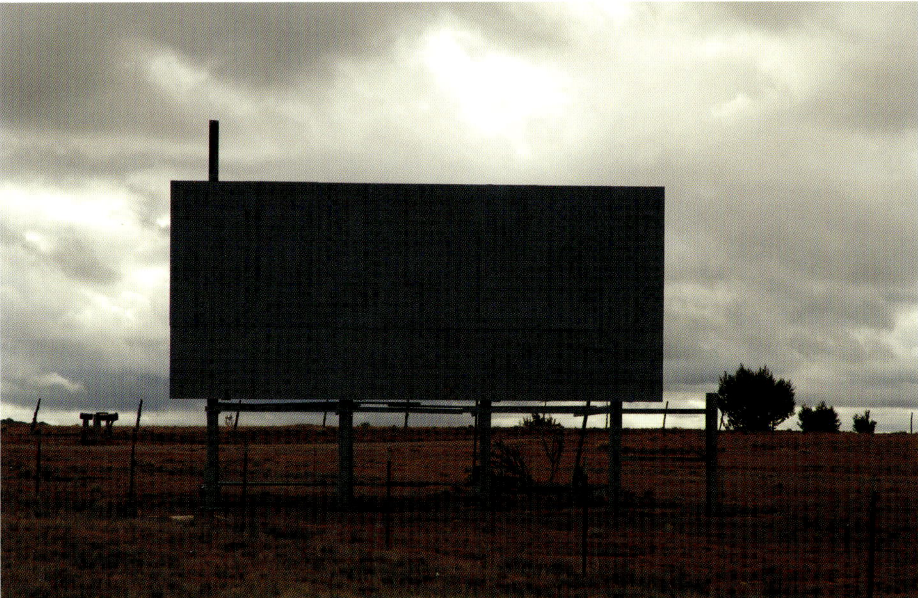

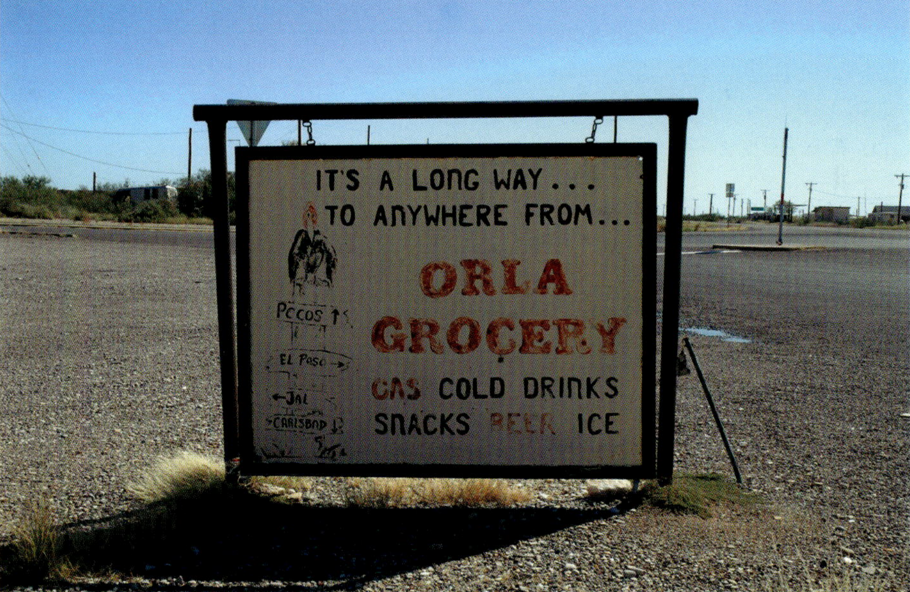
IT'S A LONG WAY...
TO ANYWHERE FROM...
ORLA
GROCERY
Pecos
El Paso
Jal
Carlsbad
GAS COLD DRINKS
SNACKS BEER ICE

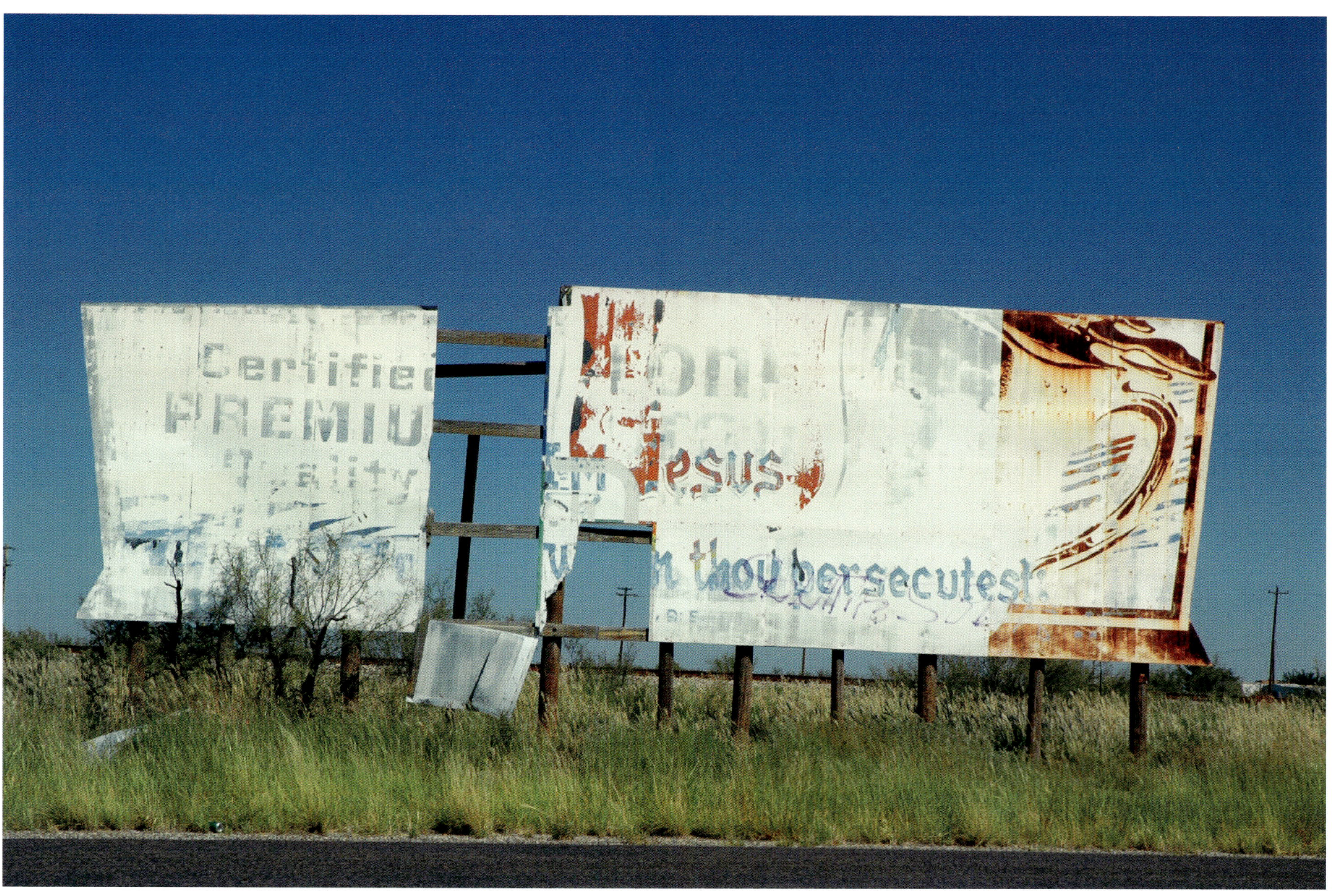
Certified
PREMIU
uality
ion
esus
n thou persecutest

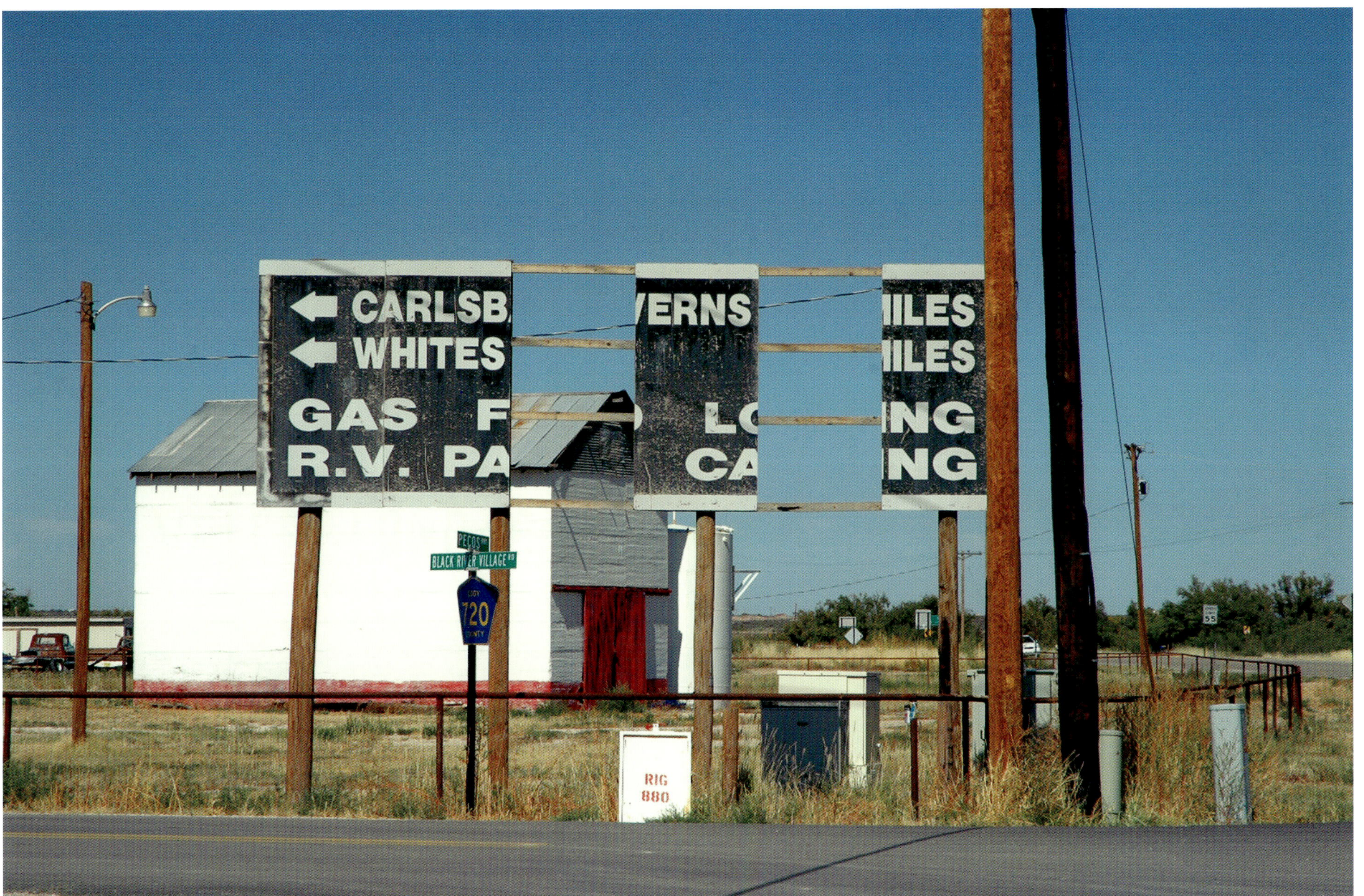
← CARLSB
← WHITES
GAS F
R.V. PA
VERNS
LO
CA
IILES
IILES
NG
NG
PECOS
BLACK RIVER VILLAGE RD
LUCY
COUNTY
720
RIG
880
55

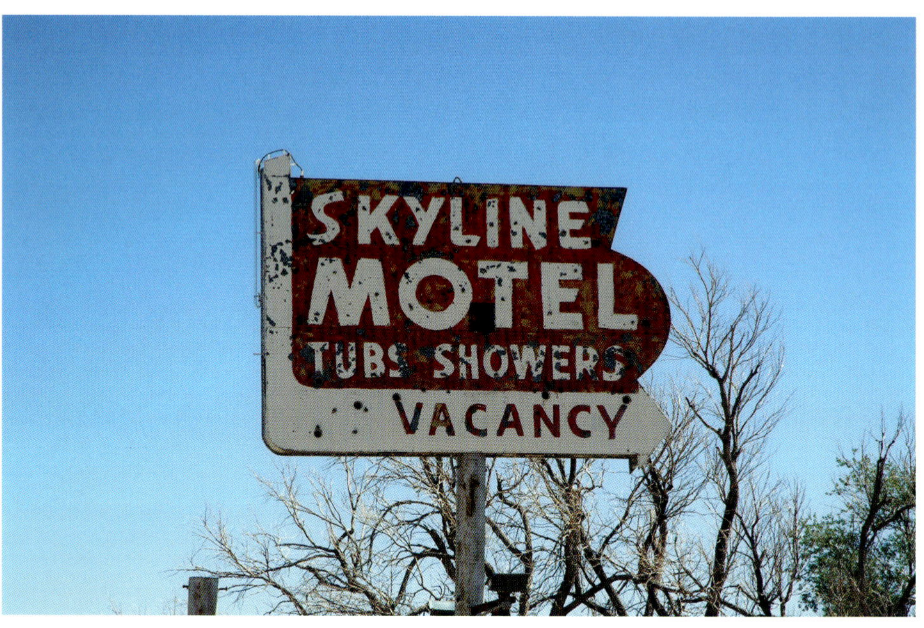

SKYLINE
MOTEL
TUBS SHOWERS
VACANCY


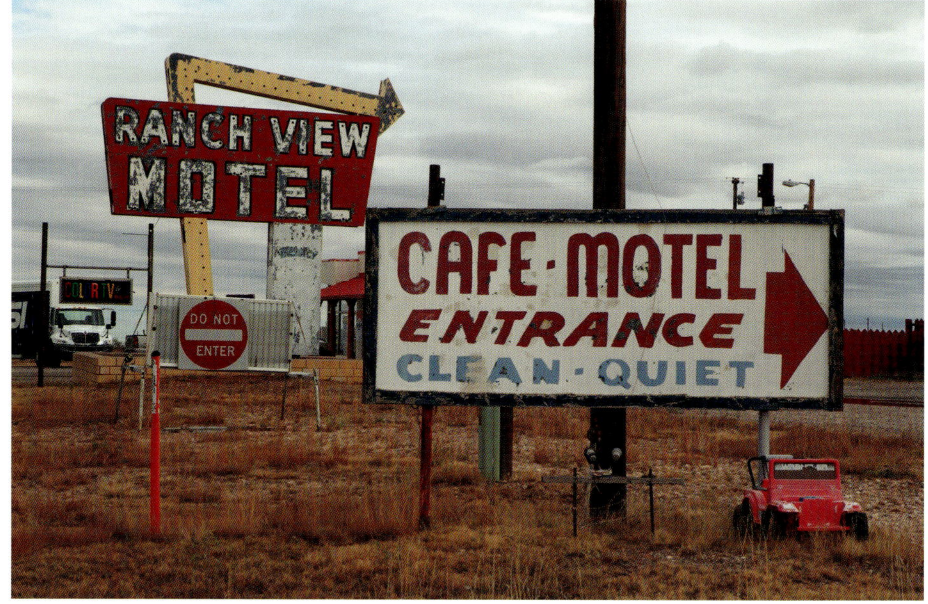

RANCH VIEW
MOTEL
DO NOT
ENTER
CAFE-MOTEL
ENTRANCE
CLEAN-QUIET

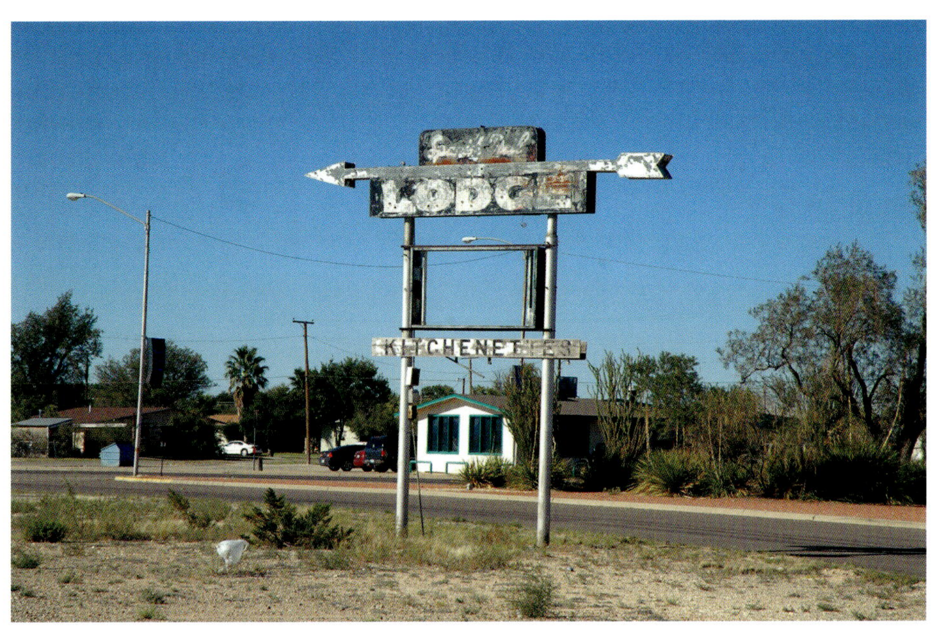

LODGE
KITCHENETTES


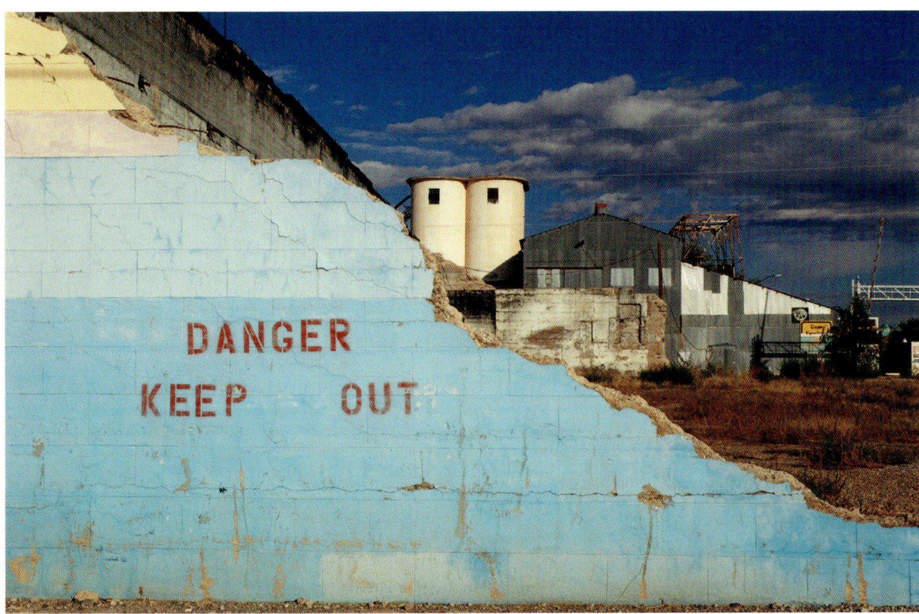

DANGER
KEEP    OUT

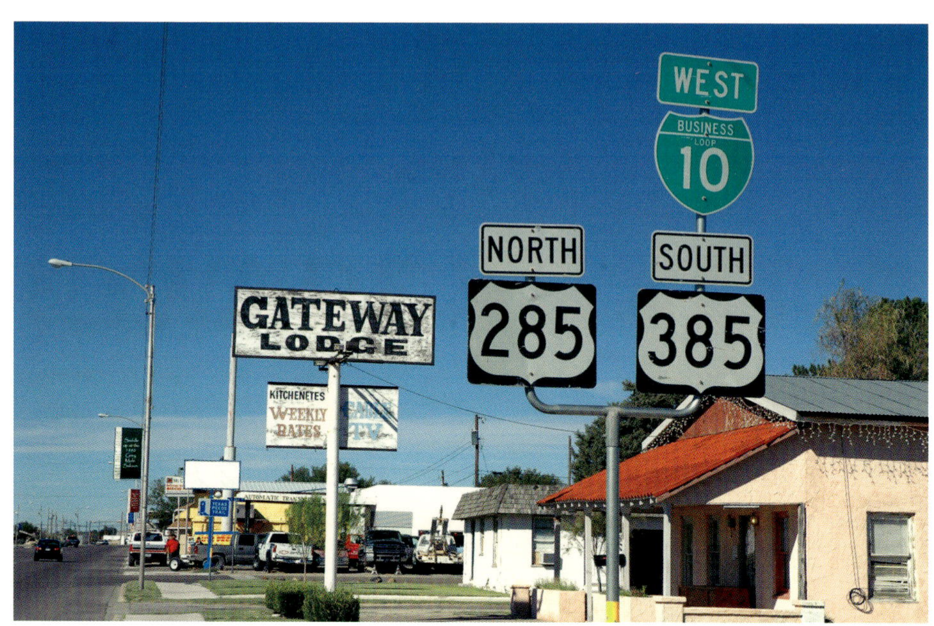

GATEWAY LODGE
KITCHENETES
WEEKLY RATES
NORTH
285
SOUTH
385
WEST
BUSINESS LOOP
10


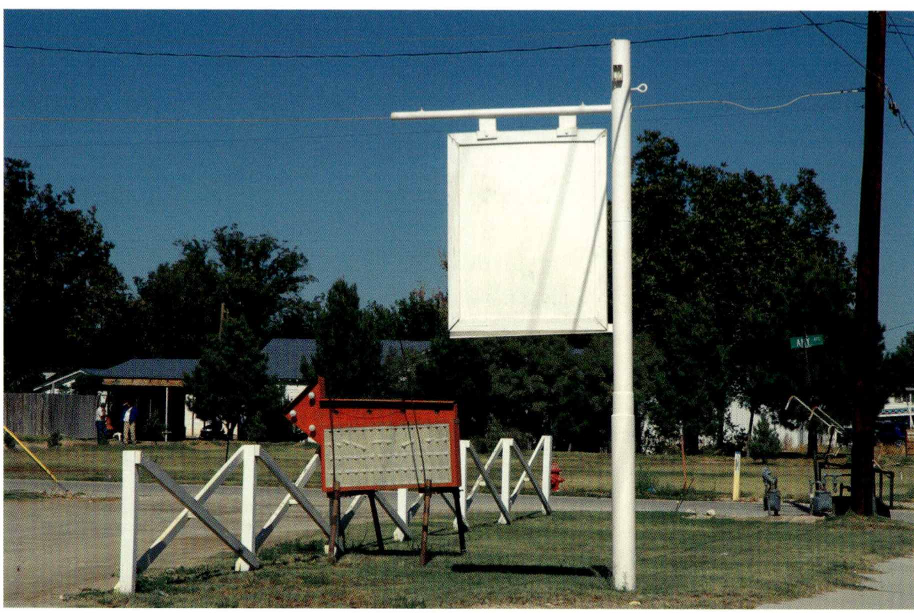

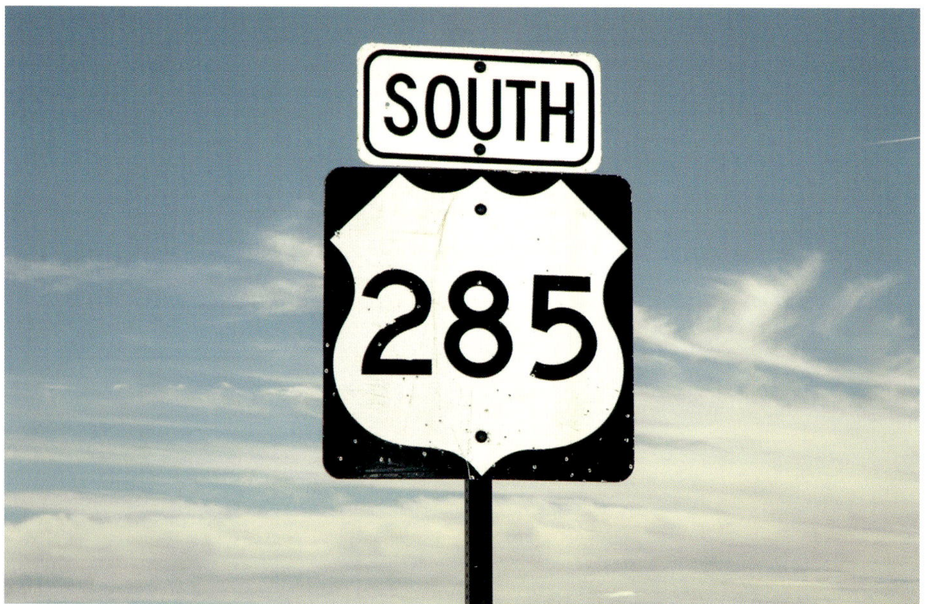

SOUTH
285


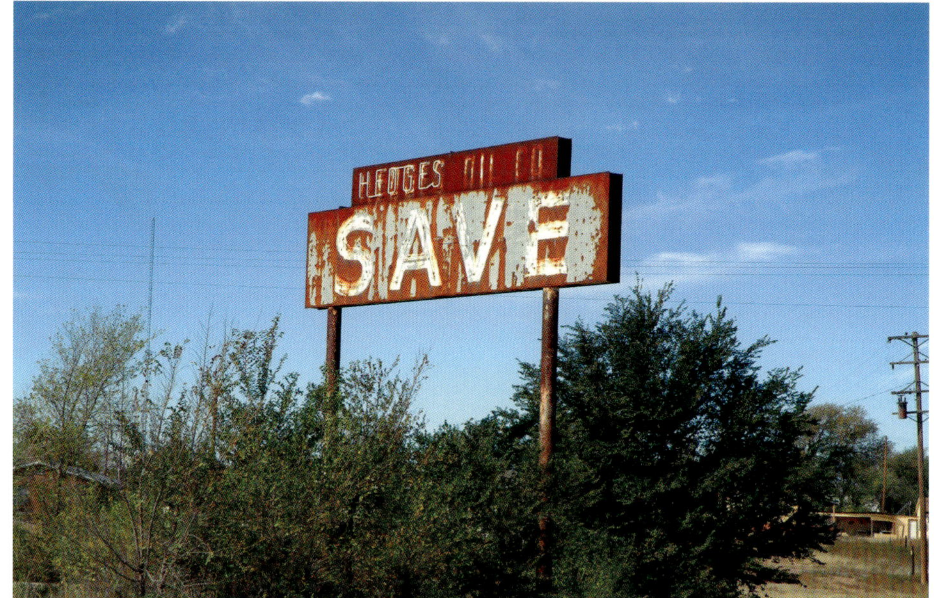

HEDGES
SAVE

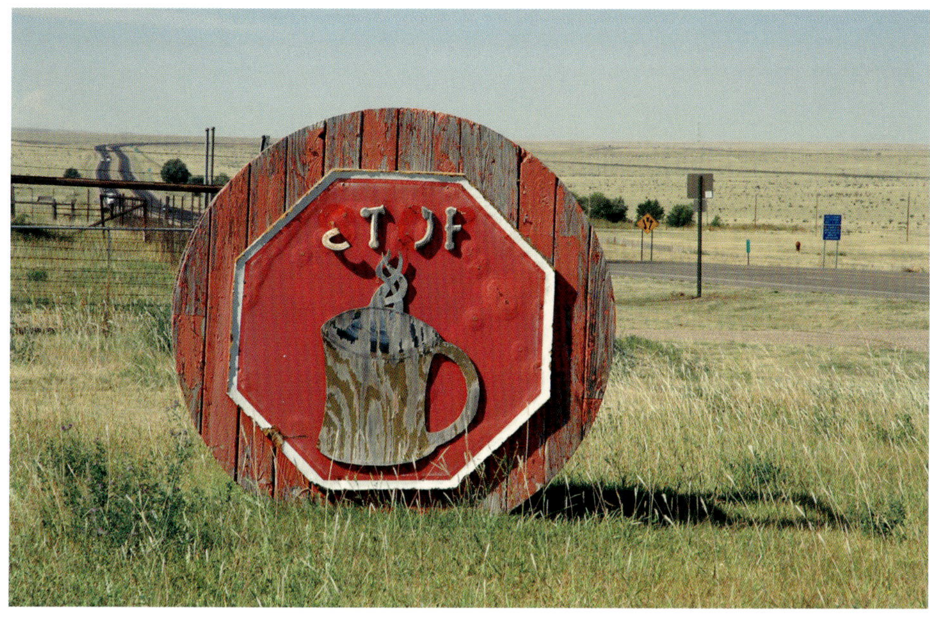

STOP


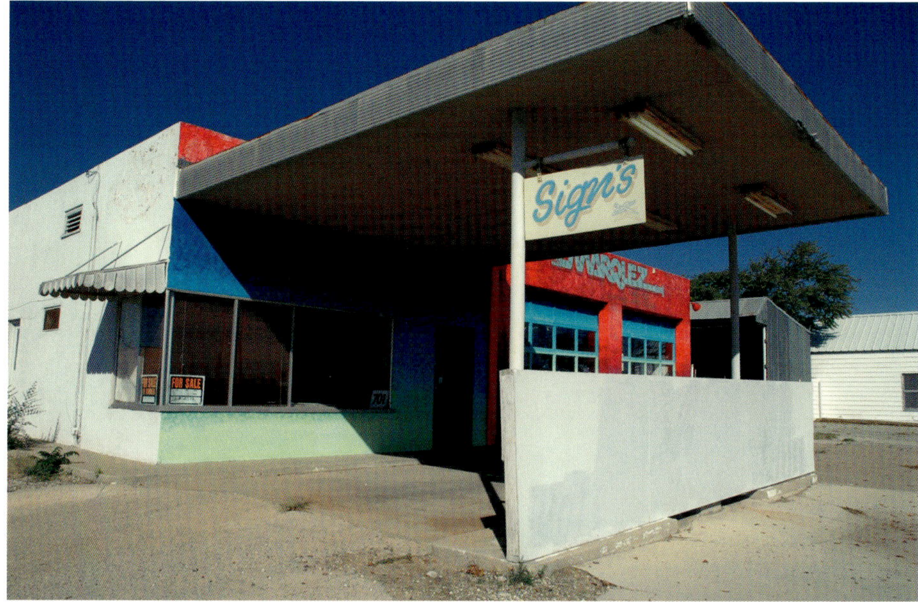

Sign's
VARQUEZ
FOR SALE


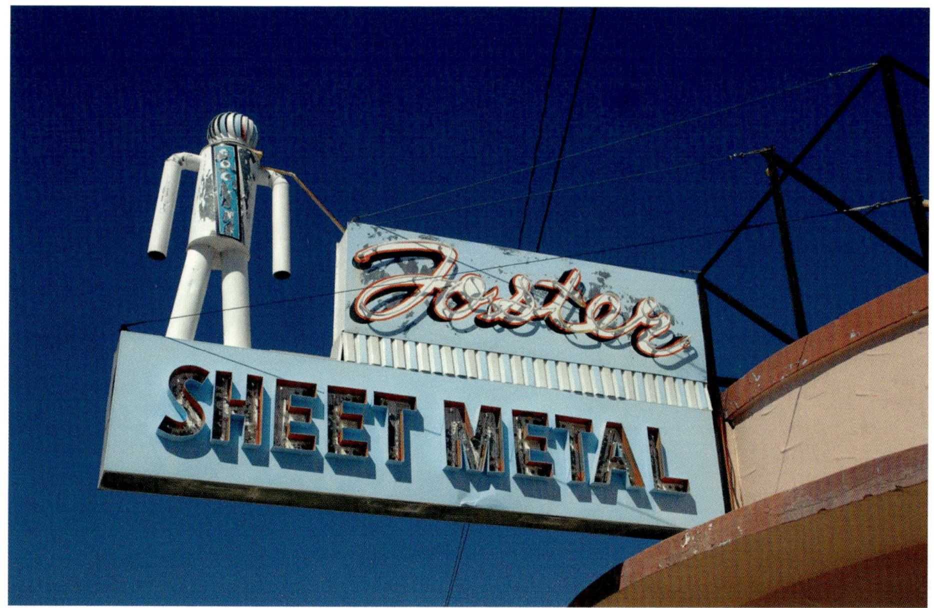

Foster
SHEET METAL


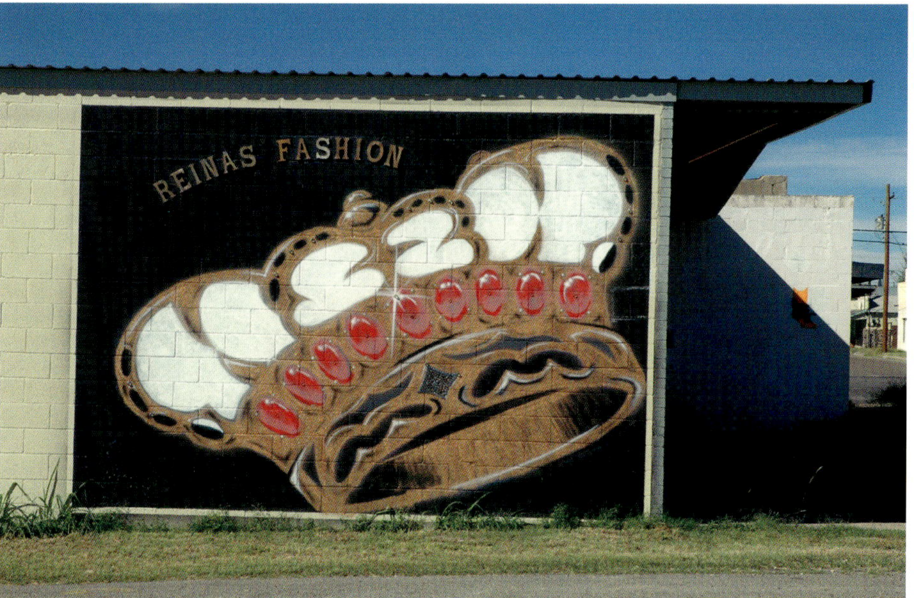

REINAS FASHION

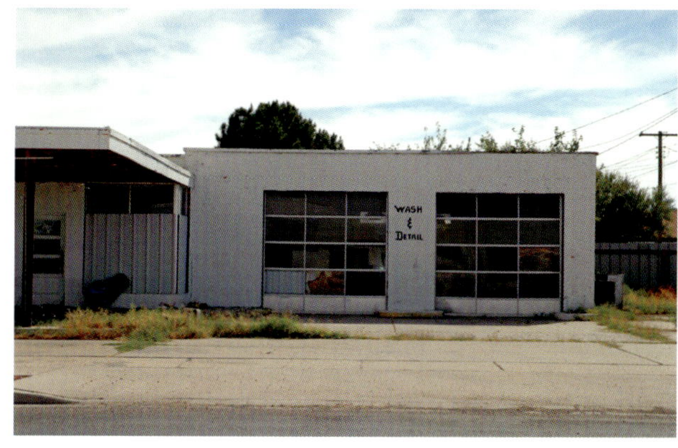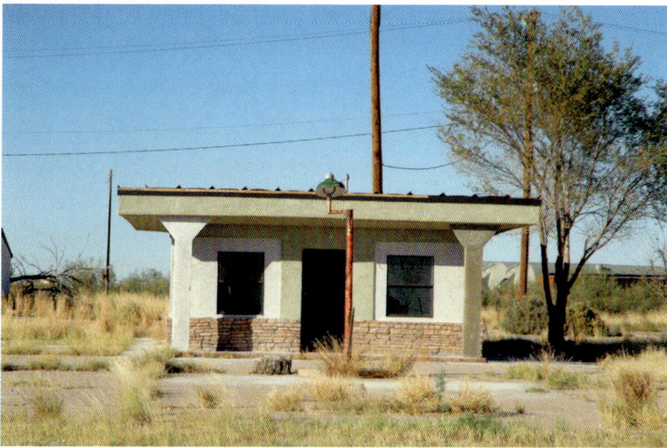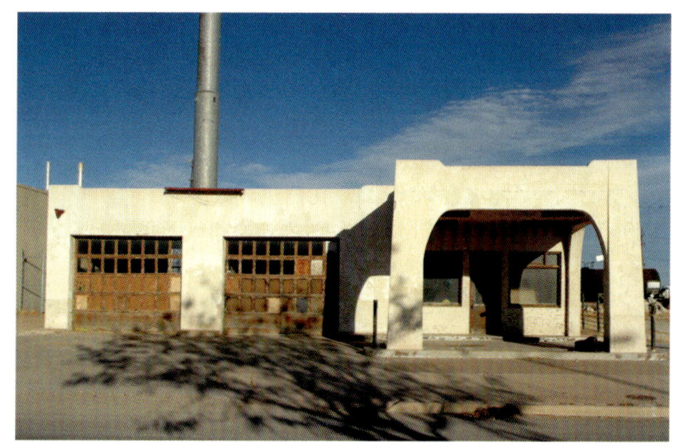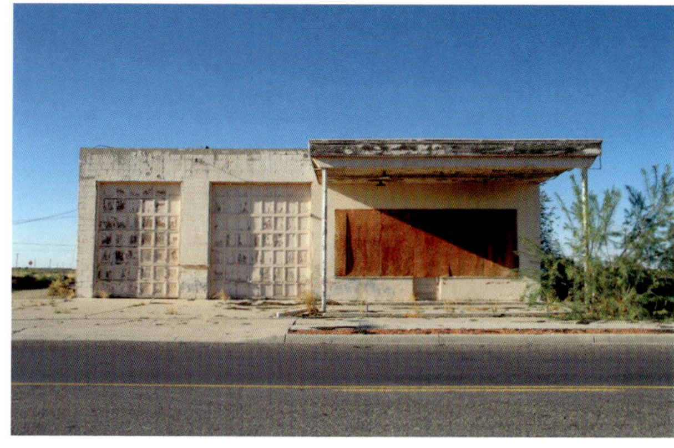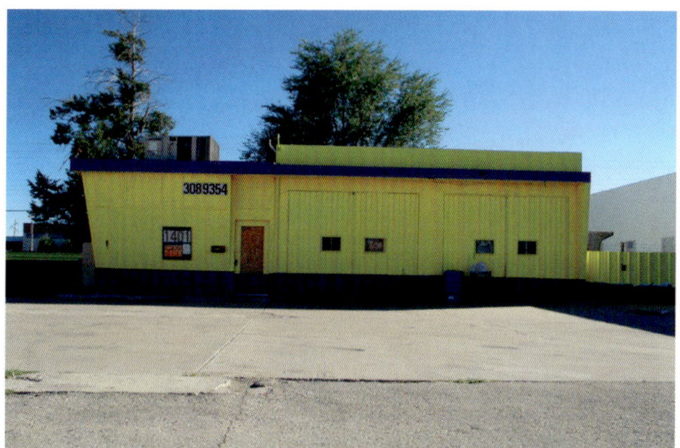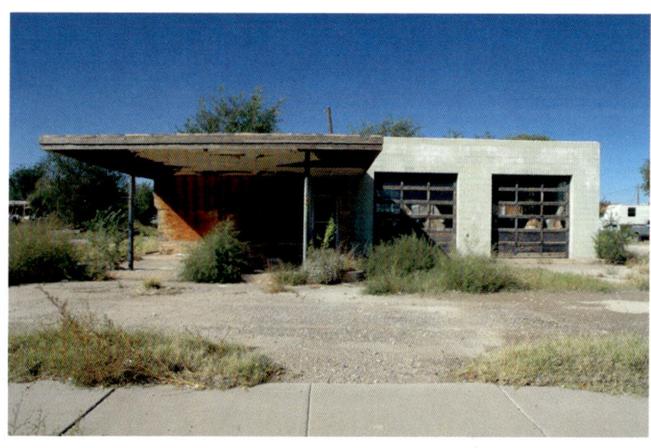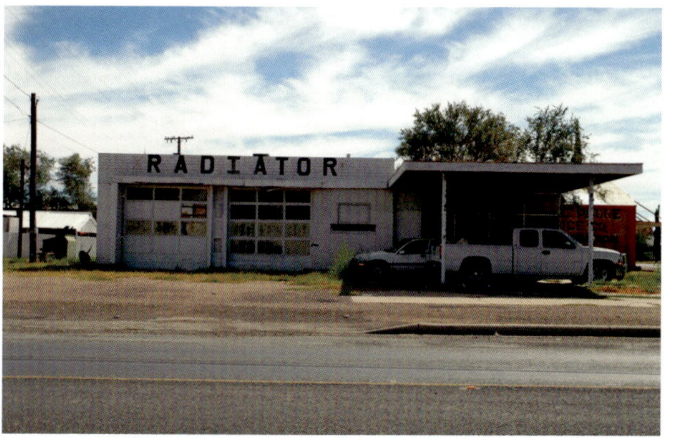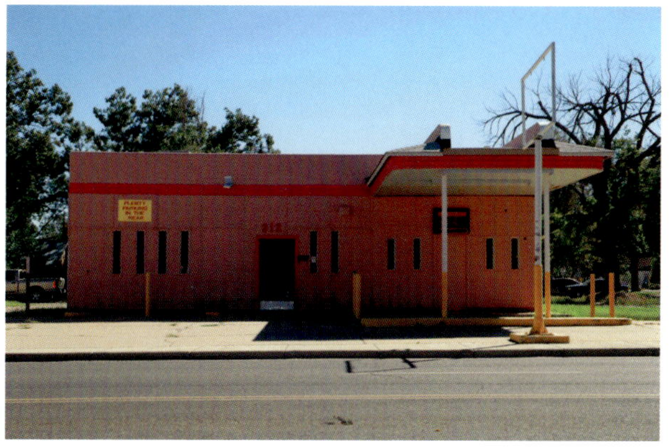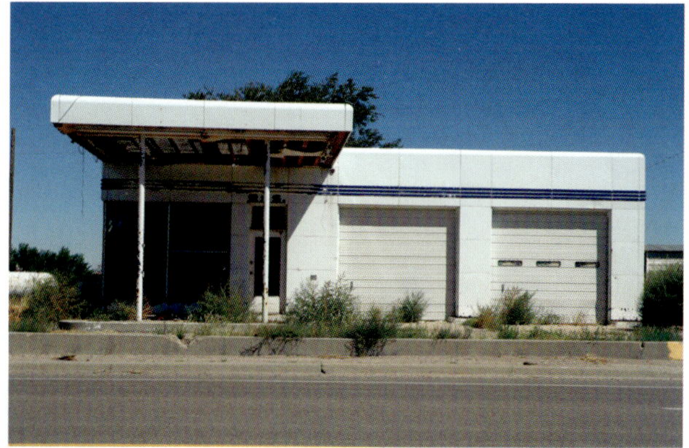

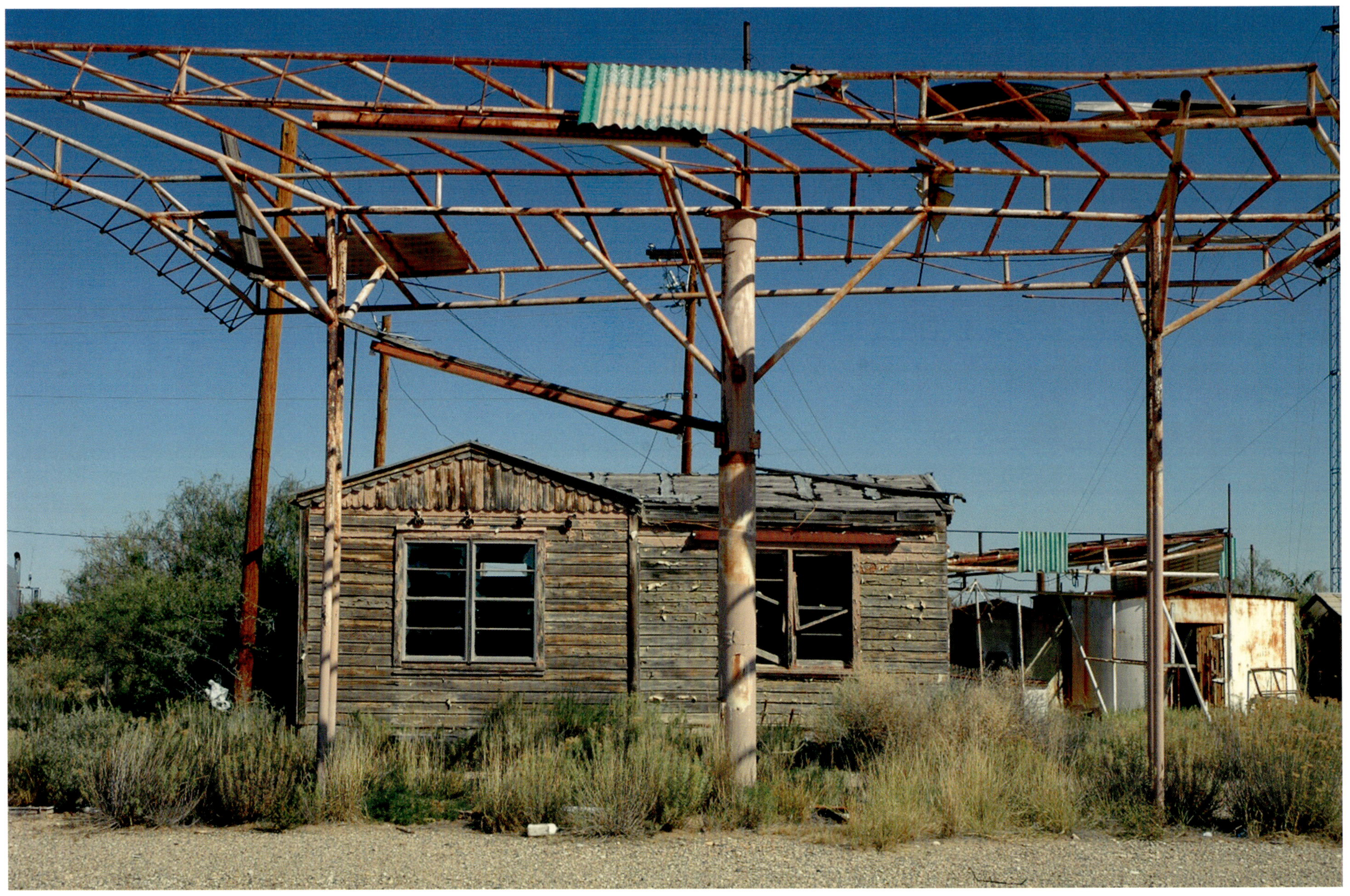

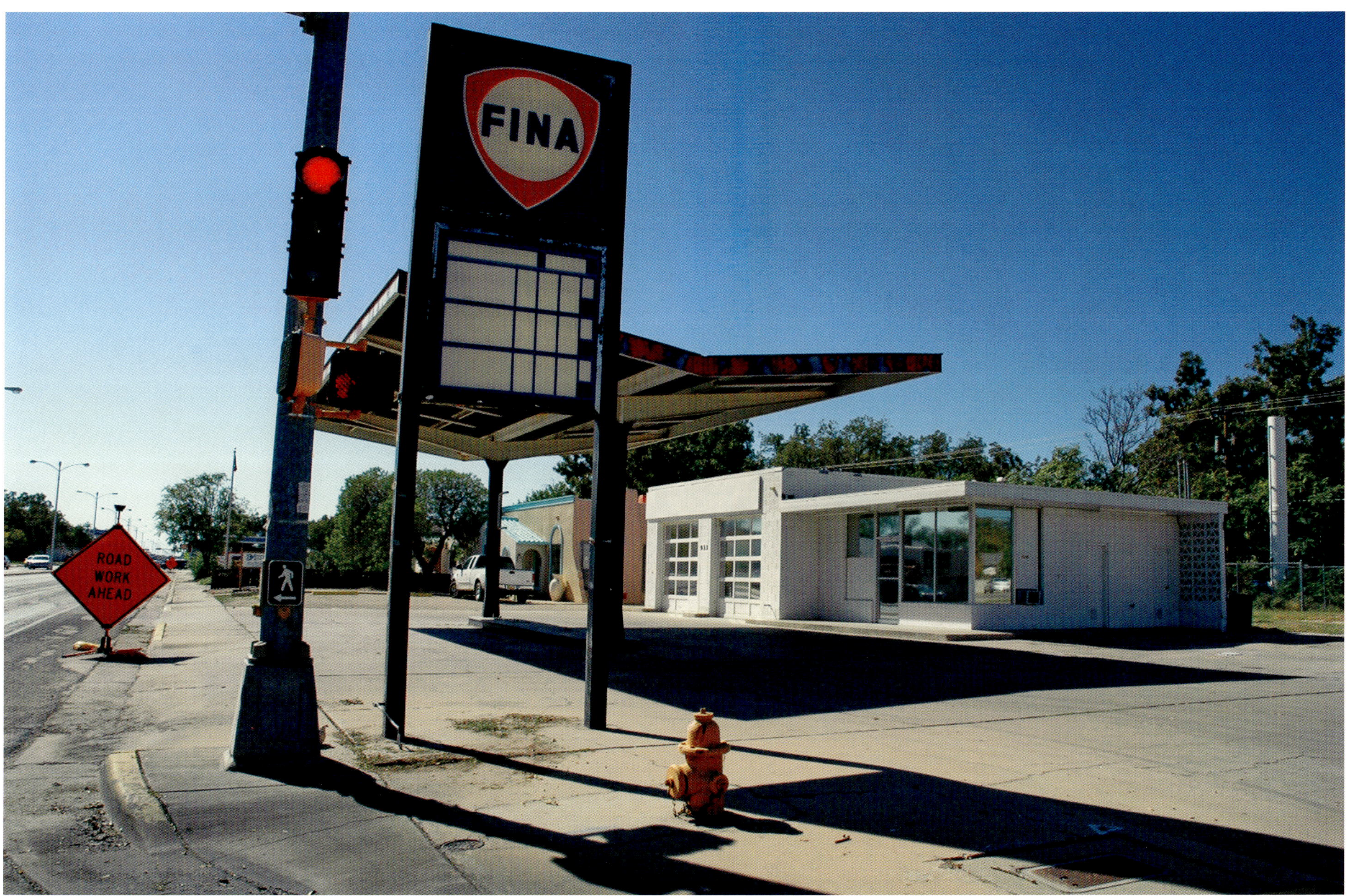

FINA
ROAD
WORK
AHEAD

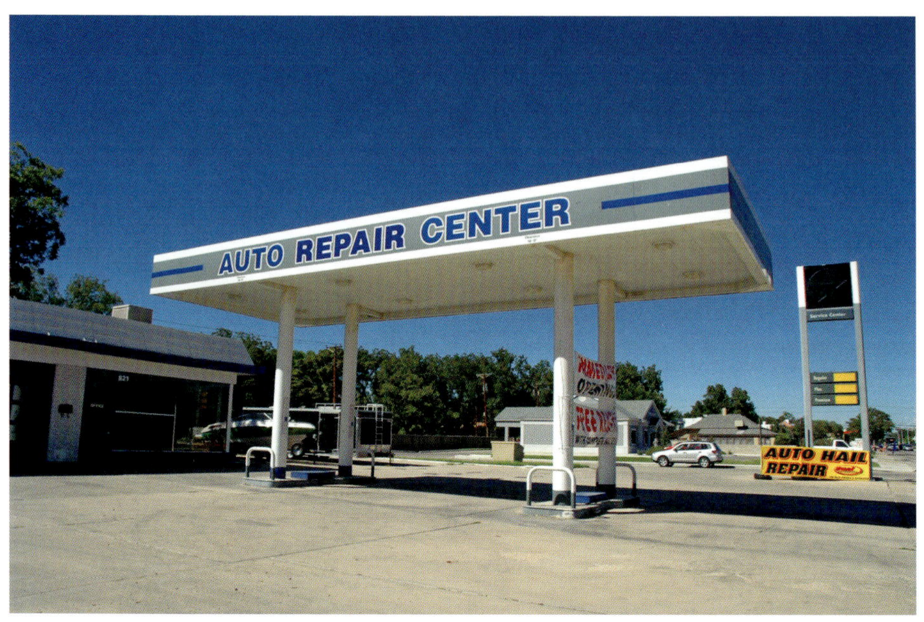
AUTO REPAIR CENTER
AUTO HAIL
REPAIR

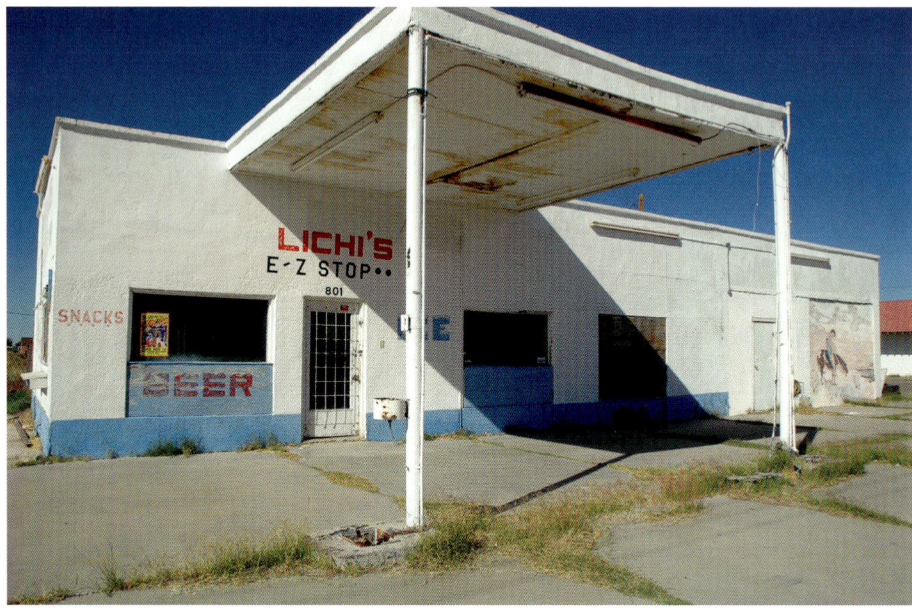
LICHI'S
E-Z STOP
801
SNACKS
BEER

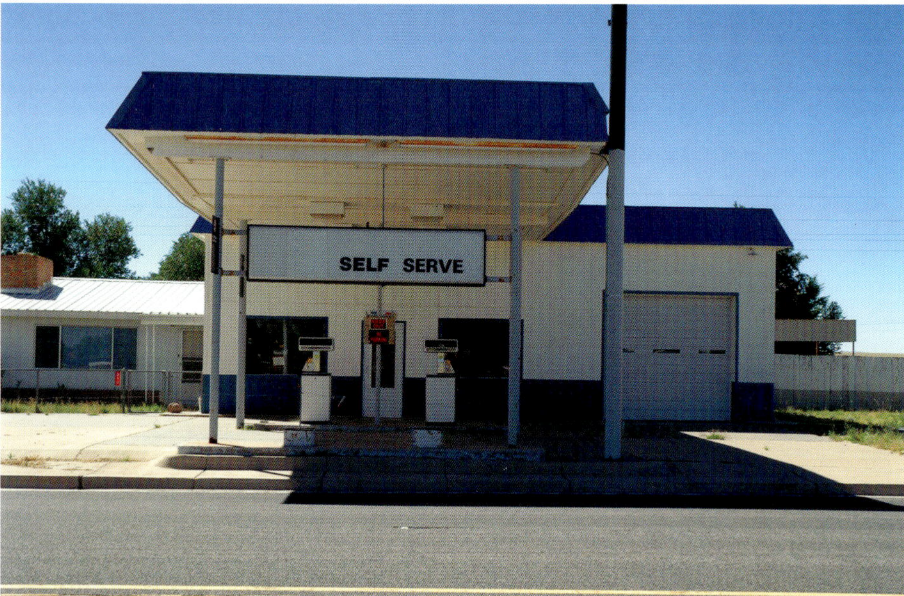
SELF SERVE

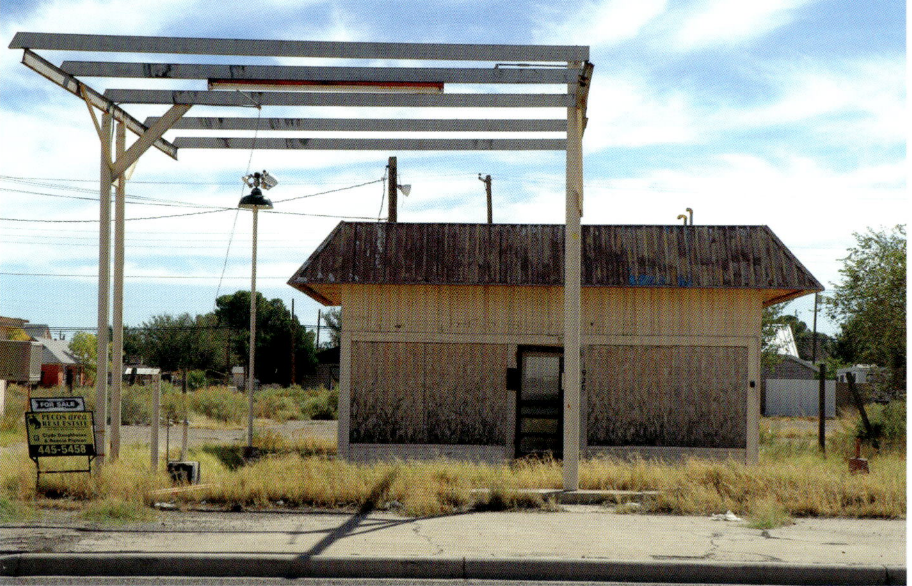
FOR SALE
PECOS AREA
REAL ESTATE
445-5458

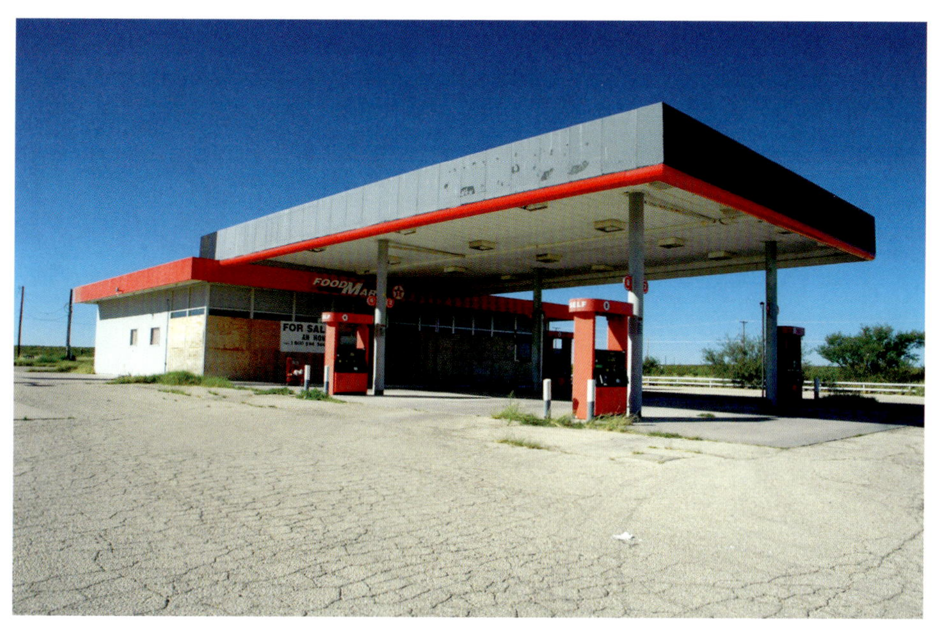

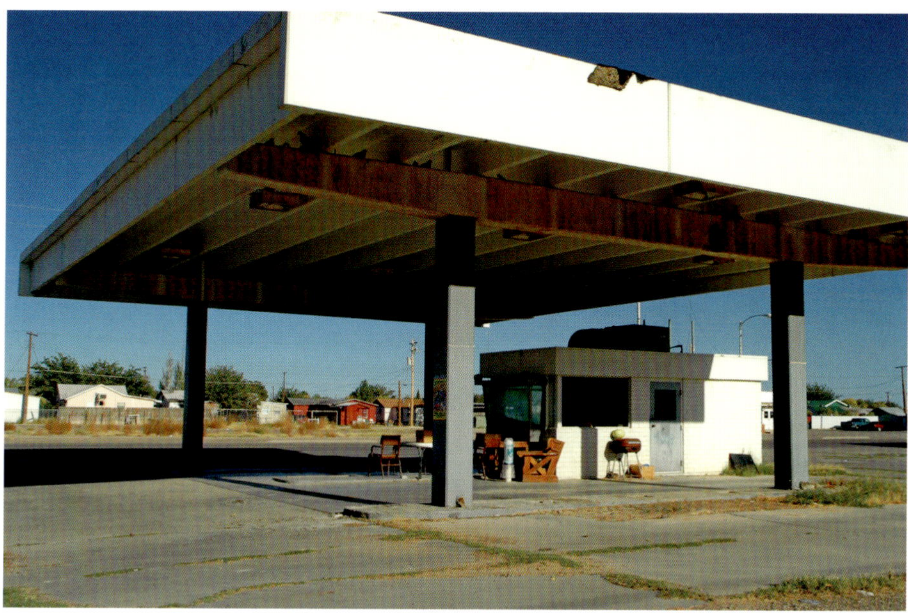

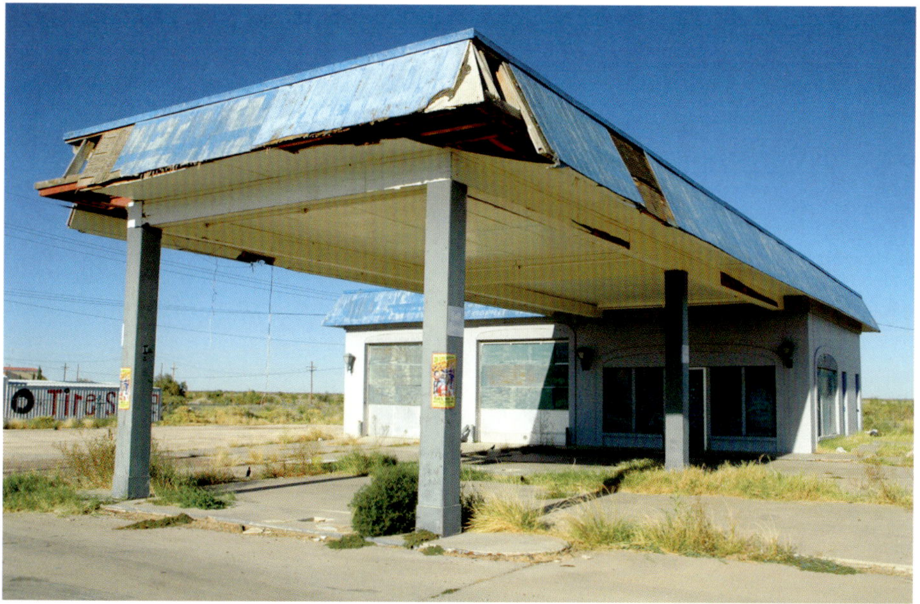

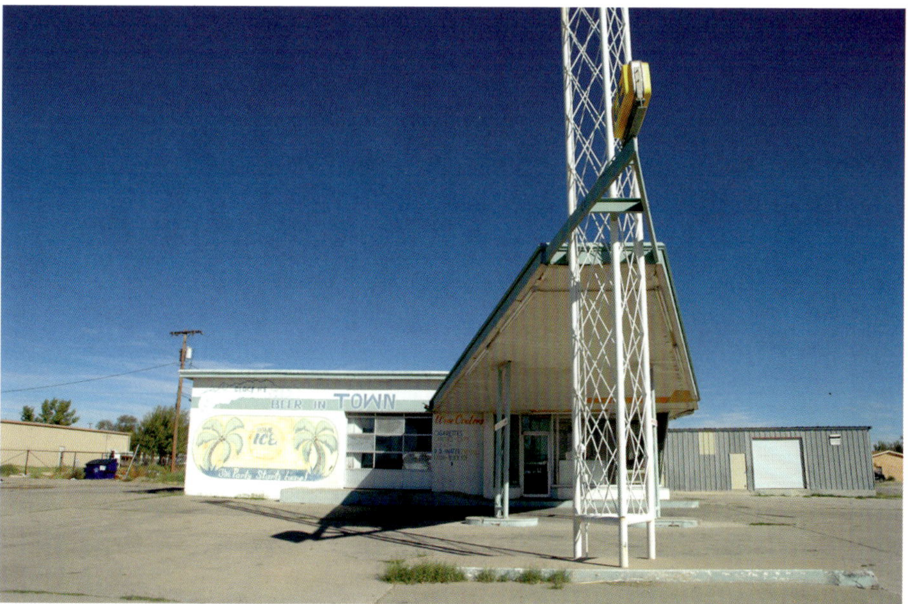

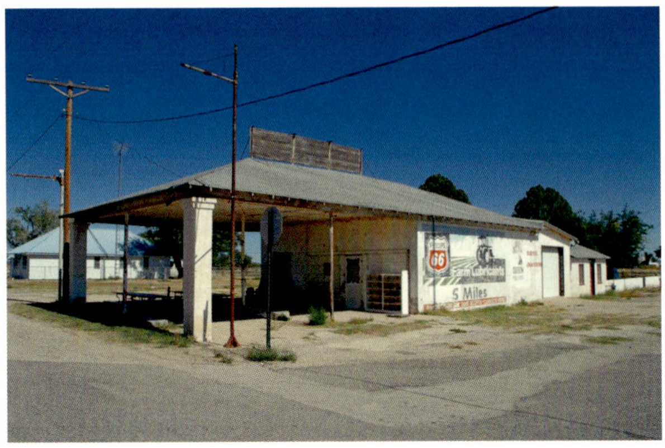

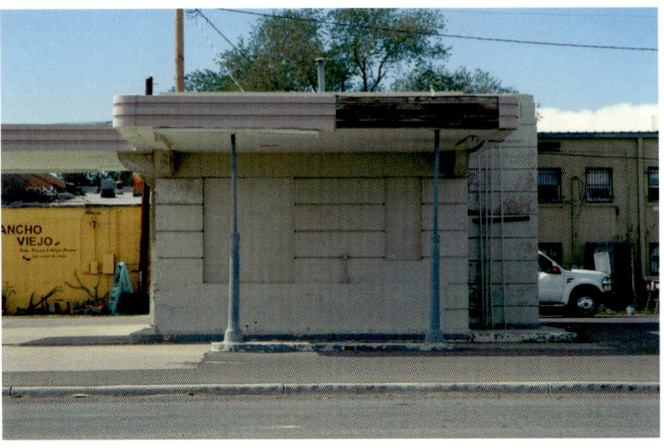

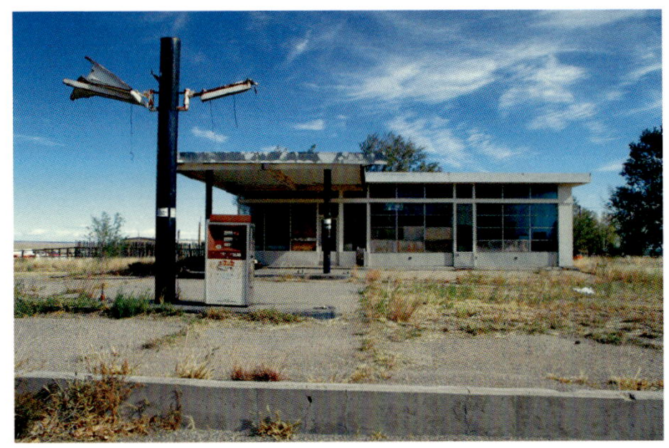

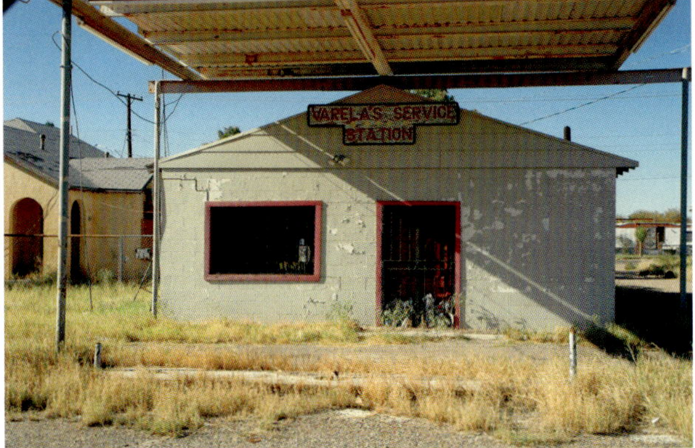

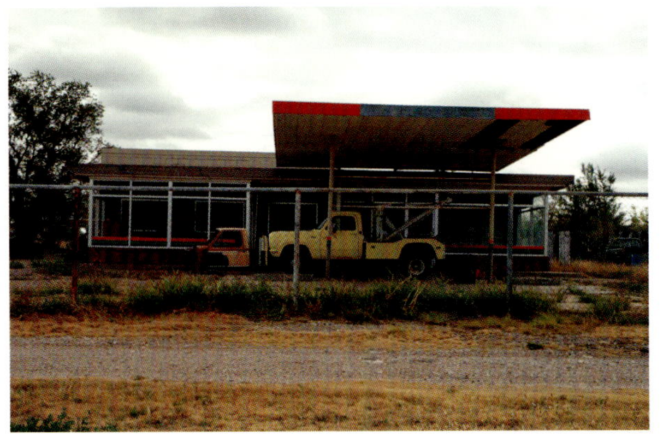

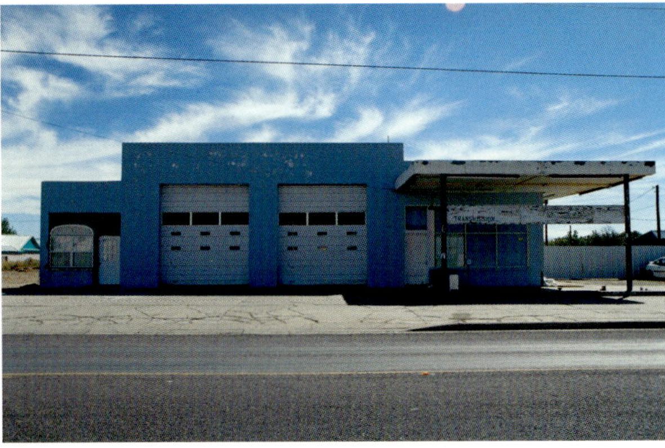

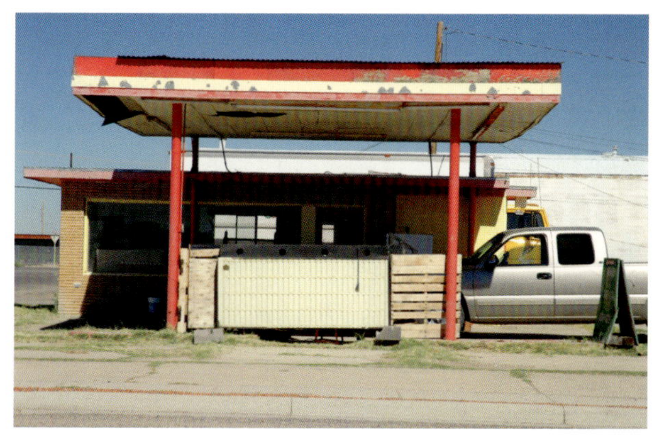
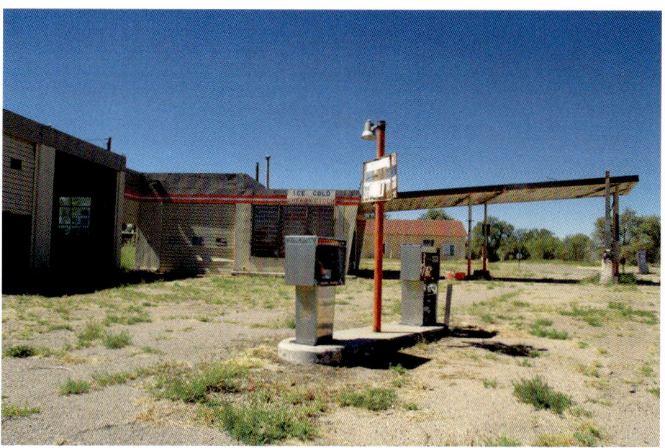
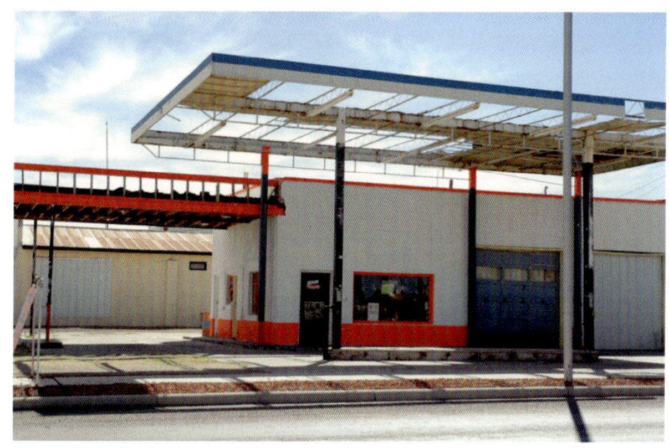
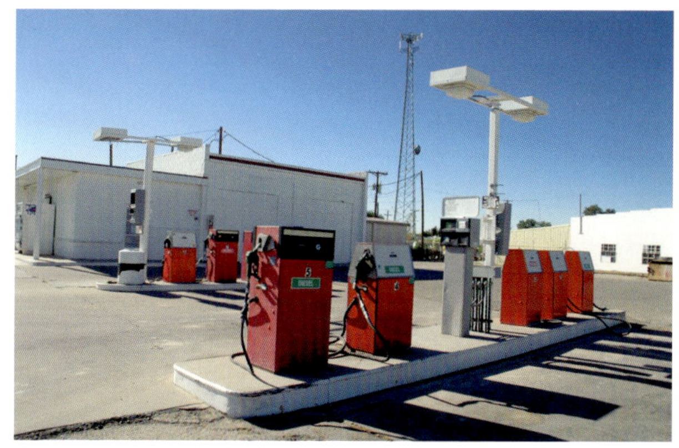
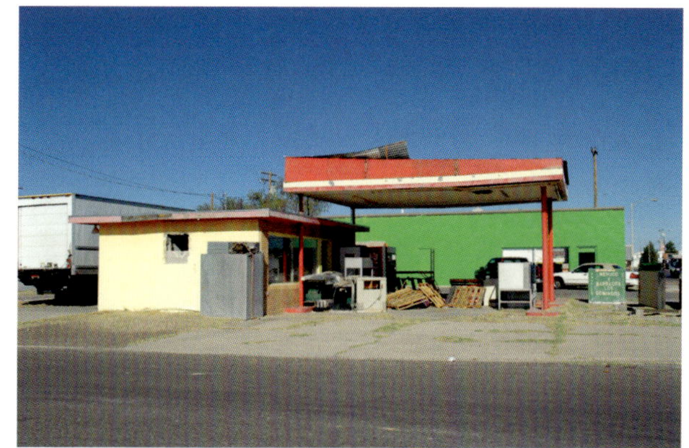
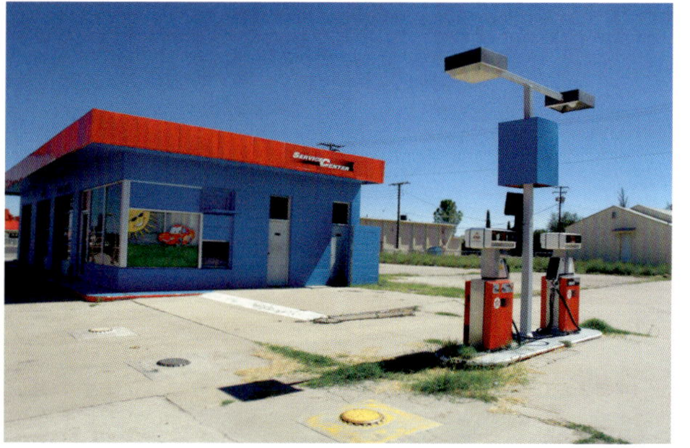
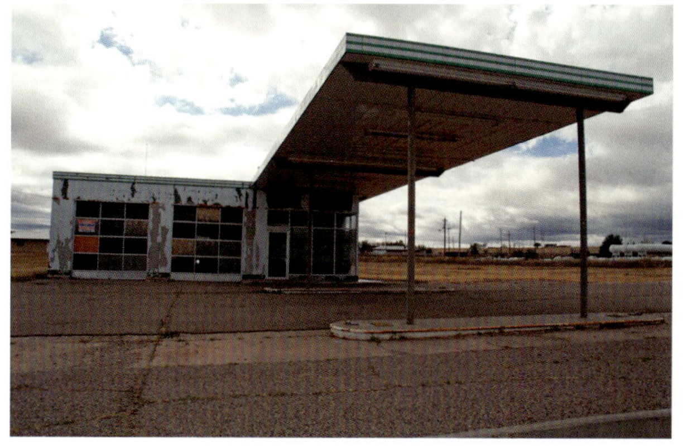
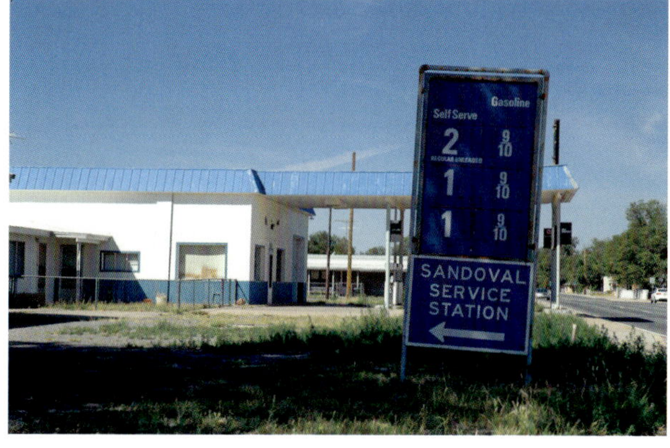

Self Serve
Gasoline
2
1
1
SANDOVAL
SERVICE
STATION

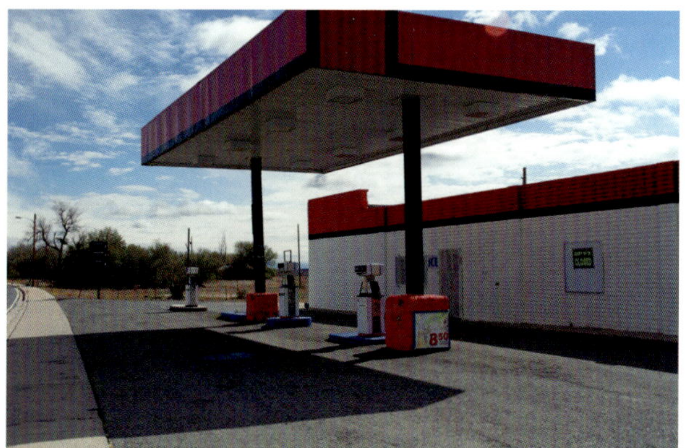

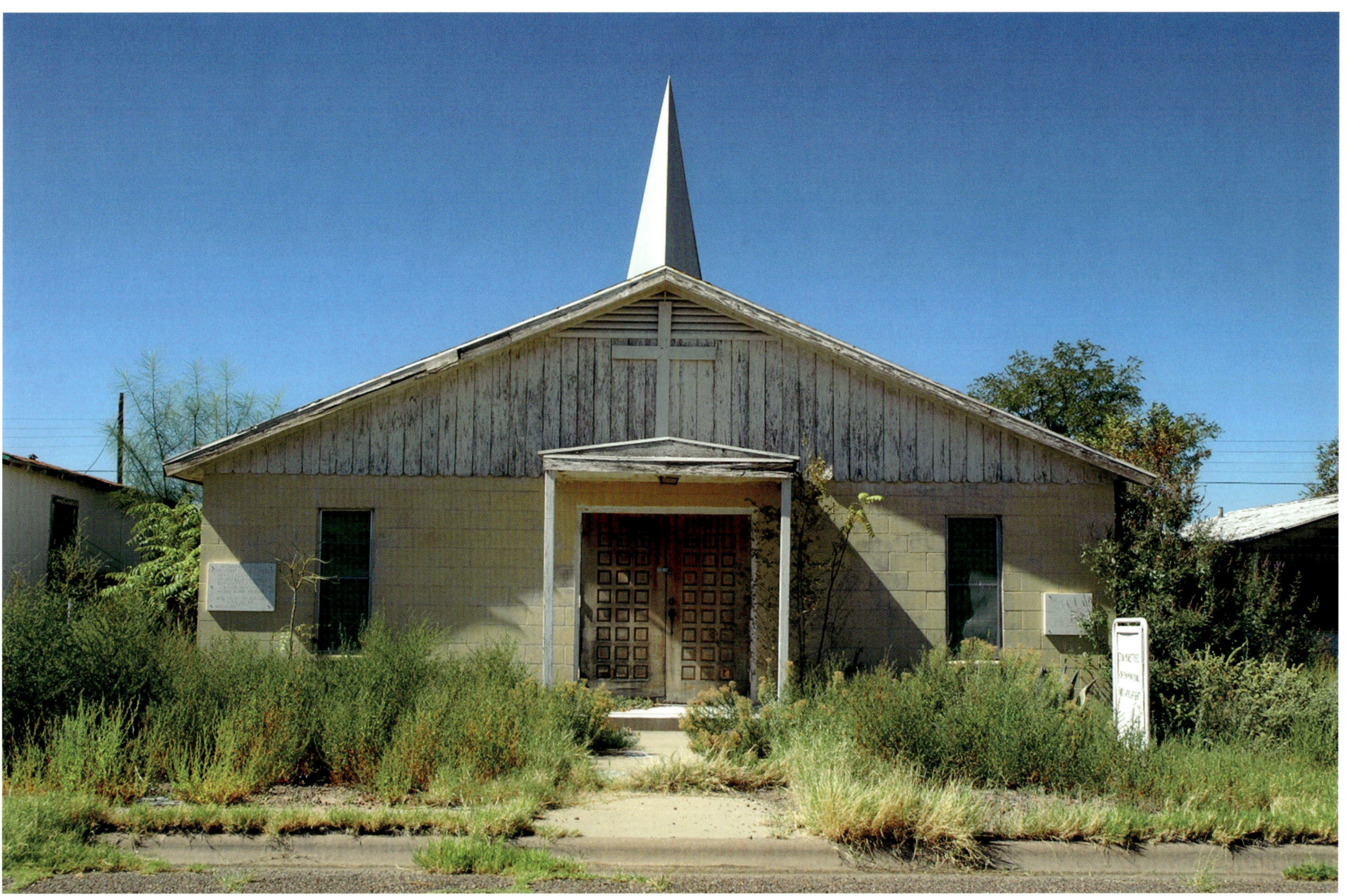

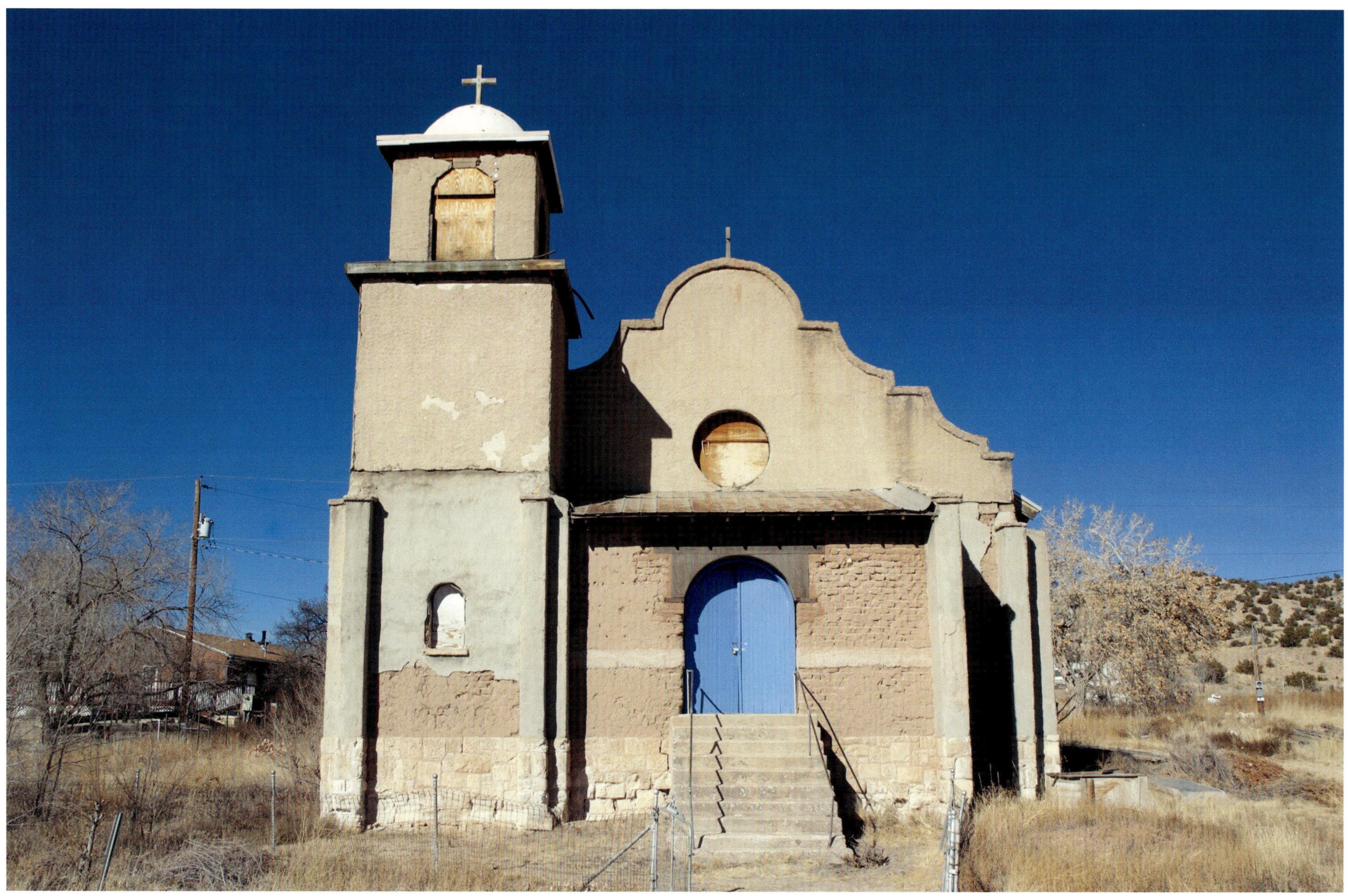

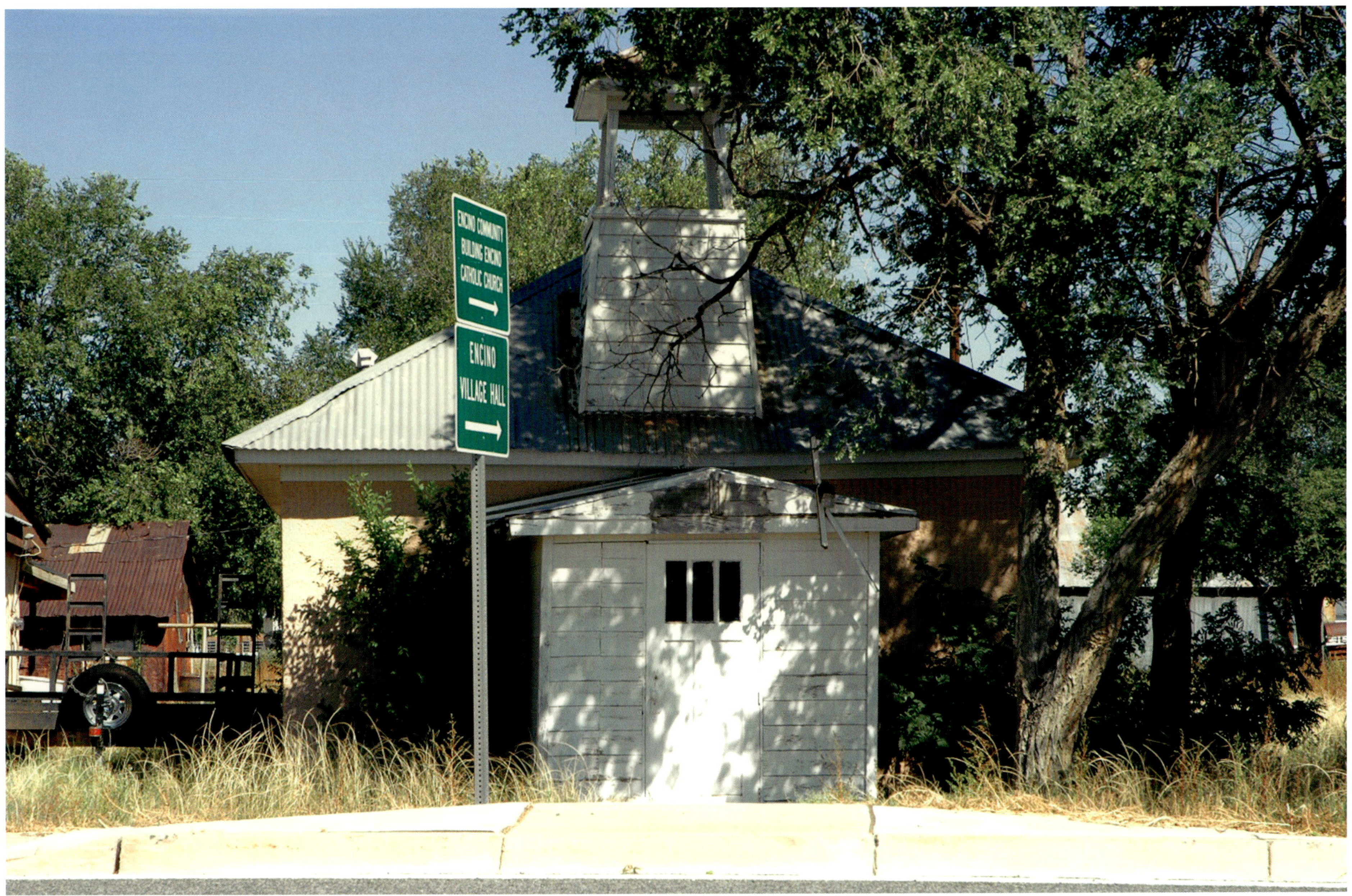
ENCINO COMMUNITY
BUILDING ENCINO
CATHOLIC CHURCH
→
ENCINO
VILLAGE HALL
→

285 Broken Dreams

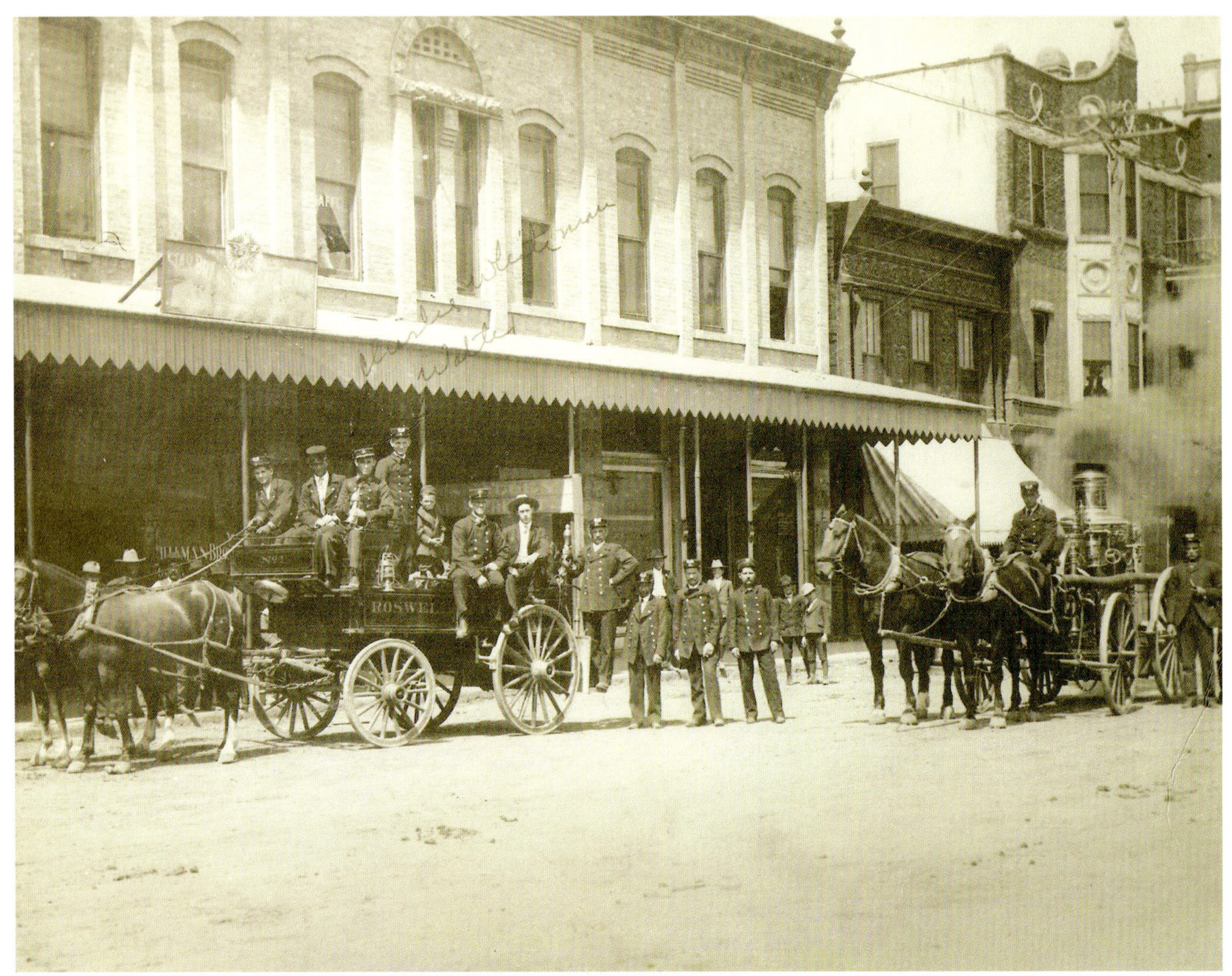

*Roswell Fire Dept., late 1890s. Courtesy HCSNM.*

## Santa Fe to Fort Stockton

BY ELVIS E. FLEMING

**IN ROGER MILLER'S** "Husbands and Wives," estranged spouses were "looking like houses where nobody lives." In this book, photographer Chris Enos presents some excellent views of "houses where nobody lives," the residences and businesses found along U.S. Route 285 between Santa Fe, New Mexico, and Fort Stockton, Texas, in the towns of a ranching region. Many of the houses and stores, clustered mainly in towns scattered across the ranchland, are very simple and cheap in construction, with a minimum of decorative flourishes. Others reflect a sense of cultivated tastes and aspirations for the finer things in life. It is known that some families of considerable means built larger houses and rented out the older houses for a time. Eventually some of these were abandoned and fell into ruin.

These abandoned structures, representing the broken dreams of people lost to history over the years, give rise to many questions, most of which will probably never be answered. Who were the people who lived in these houses and worked in these places of business? What were their dreams, and how did they become broken dreams? Why did they move here, and what developments in the broader society caused them to move away?

There have been a number of developments that brought new waves of residents to the New Mexico-Texas segment of U.S. 285 over the decades dating back to the 1860s. The spread of the Cattle Kingdom from Texas brought in cowboys, followed shortly by an influx of farmers, who cultivated the land along the Pecos River from Roswell southward and who continued to come into the Pecos Valley for several decades. Temporary employment included the building of railroads, roads, and military bases, and more permanent opportunities developed with the discovery of petroleum and in potash and sulfur mining. As these various economic activities waxed and waned, settlers tended to come and go, too.

Some of the buildings were abandoned many years ago, while others were forsaken more recently. They run the gamut of small businesses—gas stations, bars, beauty shops, retail stores, motels, cafés, and those that give no clue as to what they once were. In the case of every now-abandoned store, someone had thought, "Location, location, location. I've found a location that I believe will work!" One of the most poignant entries in the book is a photograph of a large billboard, but the billboard carries no advertisement.

U.S. Route 285

U.S. Route 285 runs 845 miles between Denver, Colorado, and Sanderson, Texas. From Denver, the route leads southwest into the Rocky Mountains and south through the San  Valley, reaching Santa Fe by way of Española. It merges with Interstate 25 at Santa Fe for a few miles, heading east through the foothills of the Sangre de Cristo Mountains, at which point U.S. 285 turns south on its own route into the plains of eastern New Mexico.

The segment of U.S. 285 on which this book focuses runs southeast from Santa Fe for approximately 450 miles, traversing the counties of Santa Fe, Torrance, Guadalupe, Lincoln, De Baca, Chaves, and Eddy in New Mexico; and Loving, Reeves, and Pecos in Texas. The principal localities along the route are Santa Fe, Clines Corners, Vaughn, Roswell, Artesia, and Carlsbad in New Mexico; and Pecos and Fort Stockton in Texas. U.S. 285 intersects I-25 at Santa Fe, I-40 at Clines Corners, I-20 at Pecos, and I-10 at Fort Stockton.

Although some of the highway is four-lane (from Clines Corners to Carlsbad, for example, because it is the route of the Waste Isolation Pilot Plant trucks hauling nuclear waste to the underground storage facility near Carlsbad), the principal use of the highway is local travel between towns. A large percentage of the traffic on U.S. 285 consists of trucking.

The present route of U.S. 285 from Denver to Sanderson was commissioned in 1936. The several segments that make up the highway were previously state-numbered roads for the most part, and the local segments each tend to have their own history. Some of the route between Vaughn and Roswell, for example, was automobile mail and stage line dating back to 1905. From Artesia south, the highway essentially follows the same mail and stage routes that were used in the 1880s and 1890s between Pecos and the various towns in New Mexico along the Pecos River.

Although each county and town has its own unique history and distinctive attributes, the areas traversed by U.S. 285 between Santa Fe and Fort Stockton have some characteristics in common, the most obvious being geography: the entire area is made up of flat, arid, wide-open spaces and includes portions of the Chihuahua Desert. Inadequate precipitation has always been a problem. The land is too arid to farm without irrigation most years, so water is precious and scarce.

A second common characteristic is the economy. The towns are few and far apart, and the land is mostly devoted to grazing cattle and/or sheep. Agriculture has always been the mainstay of the economy. Unfortunately, a key trait along U.S. 285 is poverty, which has been persistent throughout the area's history.

Demographics constitute yet another common trait. While there are great variations from one county to another, each has a mixture of Anglos and Hispanics, with limited numbers of other ethnicities. For example, Roswell, the largest of the cities in the corridor, has a diverse population that is 57 percent Anglo, 44 percent Hispanic, 2.5 percent African American, and less than 2 percent Native American and Asian American.

While there may be unique characteristics along the New Mexico–Texas segment of U.S. 285, the area undoubtedly shares in some of the national trends. In general, two of the most significant trends are the mechanization of agriculture and the urbanization of America. These developments help account for the declining rural populations and the drying up of the small towns.

### Ten Counties

The portion of U.S. 285 that is the focus of this book traverses ten counties in two states. Leaving the city of Santa Fe, U.S. 285 diverges from I-25 a few miles east and begins its journey southeast across New Mexico and Texas. The part of Santa Fe County that it crosses is some of the oldest settled land in New Mexico, harking back to the days of Spanish and Mexican land grants.

About fifteen miles south of Santa Fe, U.S. 285 intersects the first of several railroads that cross its path. In 1880, when the Santa Fe Railroad built a spur line from the main railroad to Santa Fe, the junction was named after the Lamy Land Grant. Lamy became a tourist attraction with its own Harvey House restaurant and a luxury hotel. Railroad construction and commercial development attracted workers to the area between 1880 and 1900. U.S. 285 passes about a mile west of the settlement, which is still a point of historical interest for some tourists.

South of Santa Fe County lays Torrance County, a sparsely settled ranching area where the population changes have little to do with the U.S. 285 corridor. Estancia is the county seat, and Moriarty and other villages are all in the western part of the county. The county's population reached a peak around 1940, from which it declined until the 1970s. The boom Torrance County has enjoyed since the

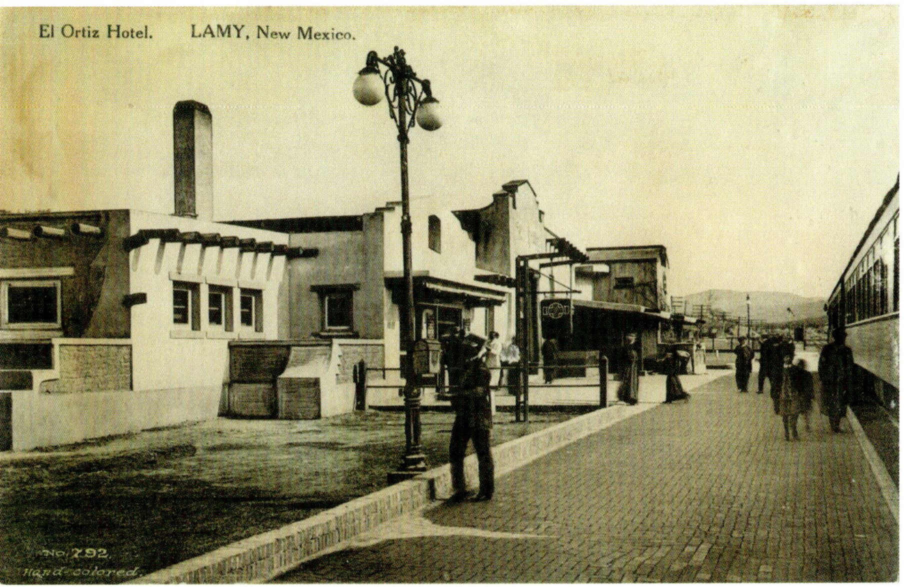


*El Ortiz Hotel, Lamy, New Mexico. Undated photograph by Ralph Twitchell, courtesy Palace of the Governors Photo Archives, (NMHM / DCA), 40312.*

1970s is a result of its proximity to Albuquerque, the largest city in the state. U.S. 285 crosses the eastern part of the county, which is largely untouched by the boom.

A famous tourist stop on I-40 where travelers can get food, fuel, and souvenirs is Clines Corners, at the intersection of U.S. 285 and I-40 in Torrance County. Roy Cline established his business in the area in 1934. After several incarnations, it finally settled at its present location. Clines Corners is not really a town, although it does have its own post office. About the only economic activity in the area that has attracted workers from time to time is the building of roads—especially Route 66 and I-40—and the maintenance of existing highways.

The little town of Encino, in the southeastern part of Torrance County, is named after the scrub oak that originally grew in the vicinity. Ranchers eliminated the shrubs by allowing sheep to overgraze the pastures. Encino has had a post office since 1904, but the town is virtually deserted, with only a few occupied residences and businesses. The Belen Cut-Off of the Santa Fe Railroad passes through Encino, and its construction attracted workers. Other developments that brought in laborers were ranching and highway building. Since ranching in the area has declined and the railroad no longer has much significance for Encino, there are many abandoned houses and businesses.

U.S. 285 next crosses into the southwest corner of Guadalupe County and passes through the town of Vaughn, far from the county seat of Santa Rosa. The county is mainly devoted to ranching, but Vaughn has always owed its existence to transportation. The site dates back to 1882, when it was a stopping place on the Jim Stinson cattle trail. Vaughn was established in the early 1900s as a Southern Pacific Railroad town. The Belen Cut-Off was completed by the Santa Fe Railroad in 1907, and Vaughn became a division point. Soon after the second railroad came through, a Harvey House hotel was built, along with a distinctive depot and a roundhouse. Like most rural villages, Vaughn has struggled to survive as residents drifted away.

Two counties that are barely touched by U.S. 285 are Lincoln and De Baca. The road cuts across the northeast corner of Lincoln County, nowhere near the county seat, Carrizozo, and the other important villages in the county: Lincoln, Capitan, and Ruidoso. The southwest corner of De Baca County is clipped by U.S. 285 a considerable distance from Fort Sumner, the county seat. Like Torrance, both of these counties historically have had sparse settlement and low population figures. The area traversed by U.S. 285 illustrates what "in the middle of nowhere" means! In the 1870s and '80s, the area between Fort Sumner and the Lincoln County towns of Lincoln and White Oaks was the stomping grounds of New Mexico's most infamous outlaw, William H. Bonney—Billy the Kid.

At times in the twentieth century, there were places along U.S. 285 that broke up the long tedious drive from Vaughn to Roswell. Refreshments and gasoline were available in Ramon, Mesa, and Chaves. The deserted houses and stores in those places speak to the problem of trying to earn a living in such remote locations.

The area along the U.S. 285 corridor in southeastern New Mexico and east toward the Texas state line has been called "Little Texas" because this part of New Mexico seemed to more closely resemble adjacent areas of Texas than it does the other parts of New Mexico, in geography, economics, religion, politics, demographics, and worldview. The people who settled in and developed southeastern New Mexico were, for the most part, Anglos from Texas and other southern states.

Proceeding south from De Baca County, before leaving New Mexico U.S. 285 crosses Chaves County and then Eddy County, both of which were created in 1889 out of giant Lincoln County. The first settlements in these two counties were established after the Civil War, with Roswell, the capital of Chaves County, being the oldest and the largest city on U.S. 285 south of Santa Fe. Roswell was started near the Hispanic settlement of Río Hondo about 1870 by Van C. Smith, who also served as the first postmaster. Carlsbad (originally named Eddy) was established in 1888 and is the county seat of Eddy County, which is named for Charles and John Eddy.

Van Smith was more interested in gambling than in town development, so it was left to Captain J. C. Lea, who arrived in 1877, to develop Roswell. Though Lea had a questionable past—he rode with W. C. Quantrill's guerrillas during the Civil War—he went on to become a true patriarch, known as the "Father of Roswell." Roswell is well known as the home of the New Mexico Military Institute, which was founded in 1891 and moved to its present location in 1898. It has long held an excellent rating as a military high school and junior college.

Cowboys, among the first workers attracted to the Pecos Valley area, were mostly associated with the cattle trails and herds brought in by cattle barons Charles Goodnight and John Chisum. There were many conflicts among early cattlemen who wanted to use the government-owned open range to start their own operations.

Several developments around 1890 stimulated population growth in Chaves and Eddy counties. Although some efforts to irrigate farmland with water from the streams were already under way, the discovery of abundant artesian water in 1890 started a boom. Land developers brought in trainloads of prospective settlers to show them the available farmlands. Many did purchase land and moved their families to the Pecos Valley from the Midwest and other areas. The development of farmland was hindered around 1893 because of flooding on the Pecos River and because of the national depression known as the Panic of 1893, which resulted from railroad and bank failures.

Both counties were organized in 1891, the same year that the Pecos Valley Railway reached Eddy from Pecos. The railroad was continued on to Roswell by 1894 and then connected with Amarillo, Texas, in 1899. It would be difficult to overstate the importance of the railroad to the economic development of the Pecos Valley.

The construction of railroads and canals, and especially the establishment of farms, brought many new workers to the area. Some of the "land and immigration" firms advertised in Europe, and as a result a group of fifty-four Swiss immigrants settled in Eddy County in 1891. They called their settlement Vaud after their home. The Swiss farmers brought in Italian immigrants to work on the farms, many of whom settled in nearby Malaga, which got its name from the sweet Spanish wine that was made from a variety of grapes that grew well in the area. Because of the influence of the Italians, the name of Vaud was changed to Florence in 1894. In 1908 it was changed to Loving in honor of Oliver Loving of the Goodnight-Loving cattle trail.

An unusual community in New Mexico was Blackdom, comprised mostly of African American homesteaders. It was established some sixteen miles south of Roswell in 1908 by Frank Boyer and his relatives and friends. Mail was delivered to Greenfield or Vocant. The Blackdom town site was platted in 1920, but the community did not

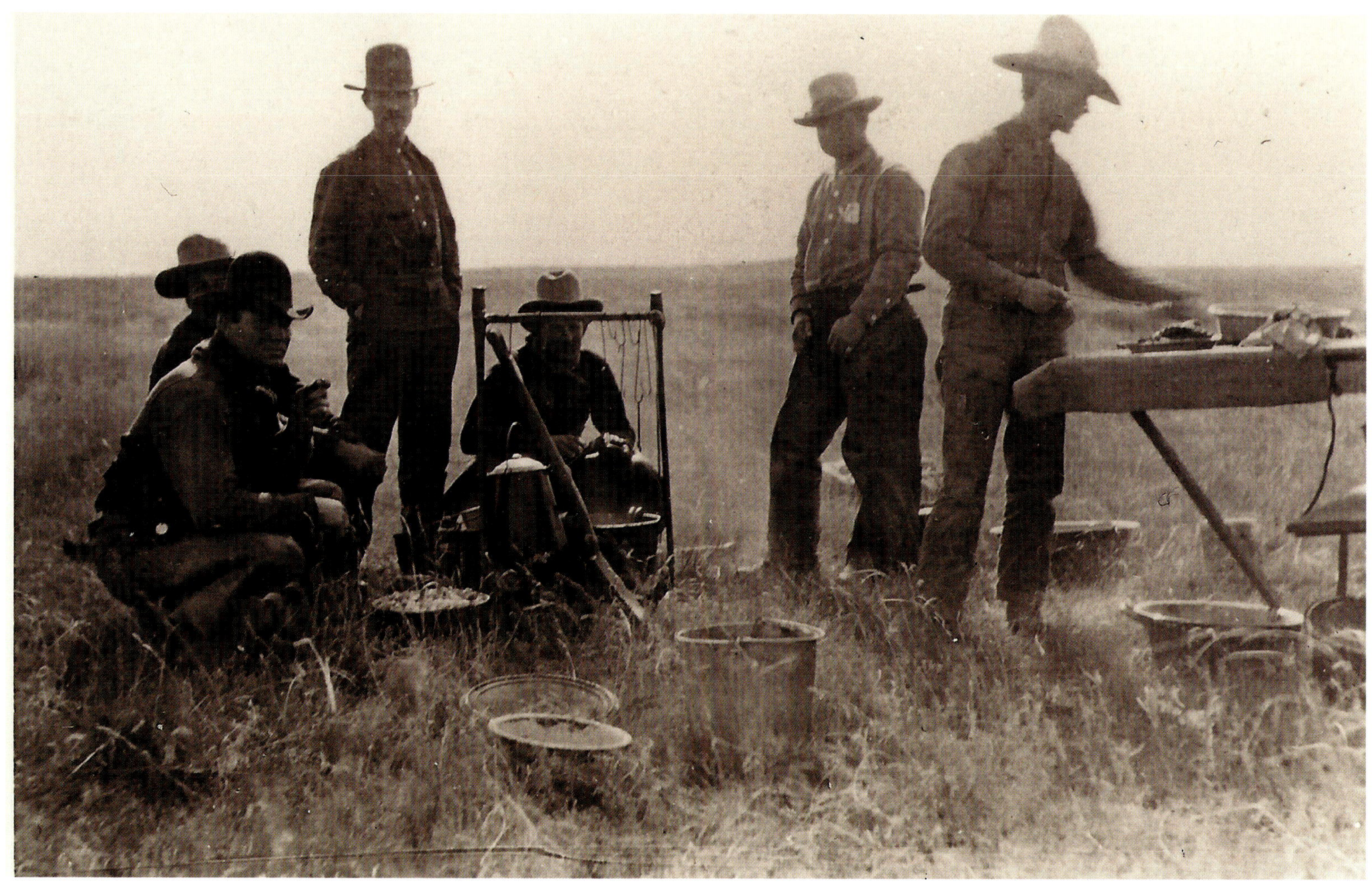

*Trail drovers break for chuck. Chaves County, ca. 1915. Courtesy HCSNM.*

thrive after that and was pretty much abandoned by 1930. It was too far west to pump irrigation water out of the artesian basin. Settlers moved to Roswell or other towns; many worked for white farmers near Dexter.

A cycle of wet years around 1909 enticed some homesteaders to northern Chaves County. The family of George Huffman, for example, managed to raise a crop of black-eyed peas, some of which they traded for pinto beans. Minor Huffman, George's son, wrote, "That winter the family had black-eyed peas one day and pinto beans the next...The dust storms played havoc with the freshly turned sod. The wind and lack of rain all contributed to the desire of the parents to get into an area where their children could get into a good school system" (Minor S. Huffman, "Huffman Family," in Elvis E. Fleming and Minor S. Huffman (eds.), *Roundup on the Pecos.* Roswell: Chaves County Historical Society, 1978, 252). The Huffmans never returned to the homestead. Their experience is a typical pattern where people try to farm the dryland prairie and are defeated by the wind and lack of precipitation.

Agriculture has always been the economic basis of Chaves County, but the nature of farming changes from time to time. Alfalfa hay was a major crop from the beginning. From about 1890 to 1930, another major cash crop was apples, until a big freeze in 1933 killed most of the remaining trees and prompted a shift to cotton production.

Going south out of Chaves County on U.S. 285, the first Eddy County town to appear is Artesia. The original route of U.S. 285 went through the Pecos Valley towns of Dexter, Hagerman, and Lake Arthur—all in Chaves County along the Pecos River. Certainly those towns have had their share

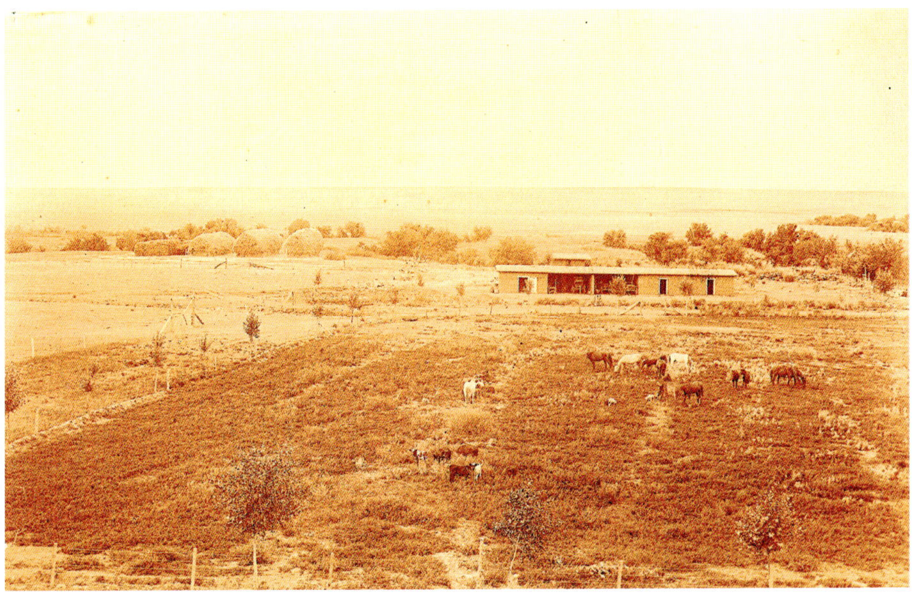

*Diamond A Ranch, Chaves County. Courtesy HCSNM.*

of broken dreams, especially Lake Arthur. In 1967 a new route laid out for U.S. 285 proceeds almost directly across the prairie from Roswell to Artesia, bypassing the smaller towns. Therefore, most of the few abandoned structures on U.S. 285 between Roswell and Artesia are relatively new in comparison to other segments.

Artesia is the second-largest town in Eddy County. It started as a cow camp on John Chisum's range and became a stage station and

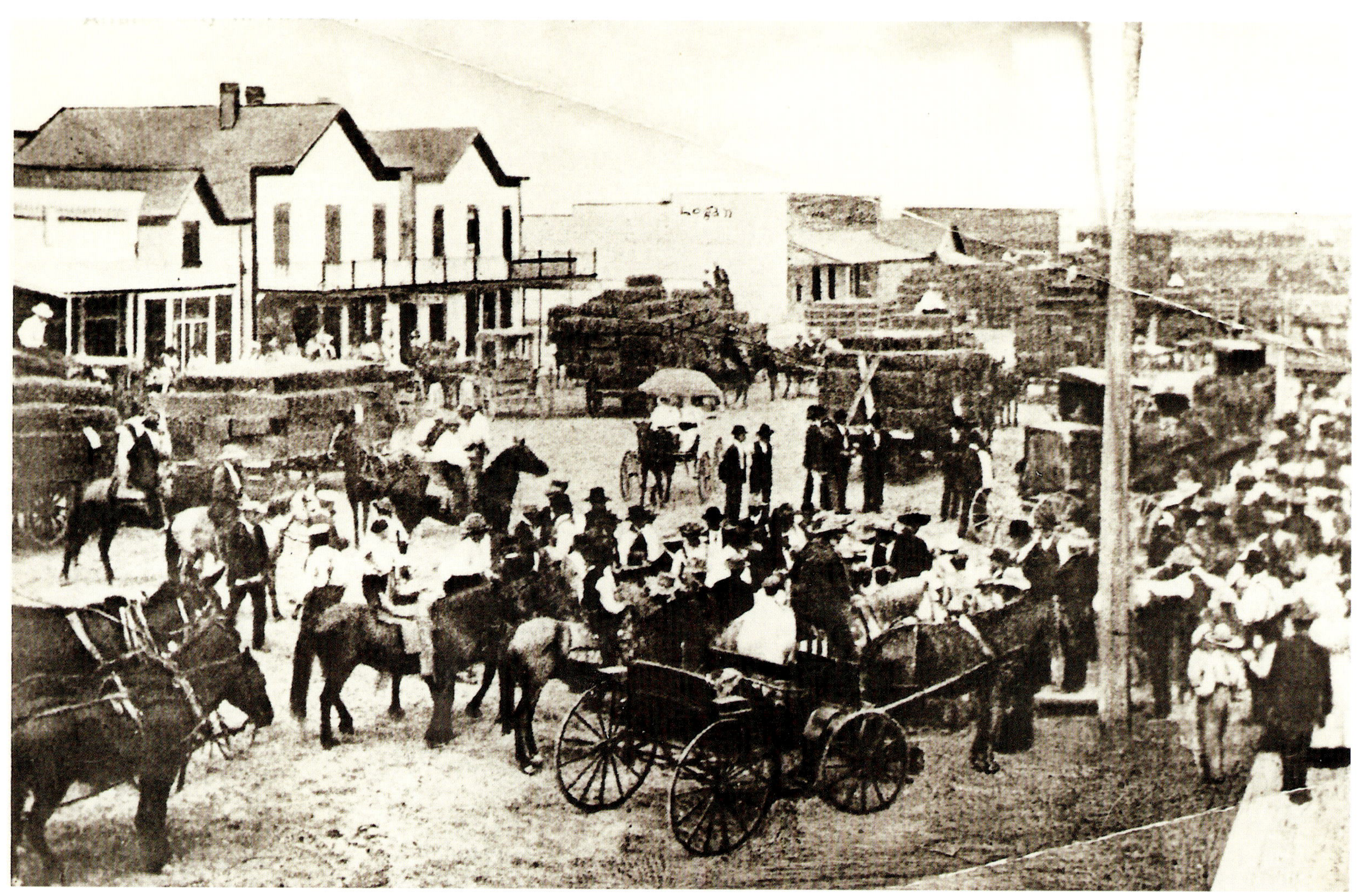

*Alfalfa Day in Artesia, early 1900s. Courtesy HCSNM.*

*Carlsbad Spring, 1920. Photograph by Annie Laurie Snorf, courtesy HCSNM.*

then a stopping place on the Pecos Valley Railway while it was being built to Roswell from Eddy in 1894. The town is a center for farms and ranches, but its principal economic activity has been petroleum since about 1924. After several attempts the first producing well was brought in, followed by numerous discoveries of oil fields and wells, especially in Eddy County but also in Chaves, Loving, Reeves, and Pecos counties. Oil production was then accompanied by the construction of pipelines and refineries. The economic boom brought into southeastern New Mexico by the oil industry attracted many workers over the years.

South of Artesia is the city of Carlsbad, which started as Eddy in 1888. It took the coming of the railroad in 1891 to get the town going. It was chosen as the county seat of Eddy County when it was organized in 1891. In 1899 the name of the town was changed to Carlsbad because a spring near the town resembled one in Europe with that name. In the early days, Carlsbad had suburbs that seemed to fit into the stereotype of the Wild West: Seven Rivers, Phoenix, and Lone Wolf. Carlsbad enjoyed all the same stimuli to its economy as Roswell and Artesia, such as ranching, farming, the building of irrigation canals, and the development of the oil industry. In addition, Carlsbad benefited from the beginning of potash mining in 1931—an activity that is still a major factor in the local economy. It became a tourist destination in 1930 when the bat cave southwest of town became Carlsbad Caverns National Park.

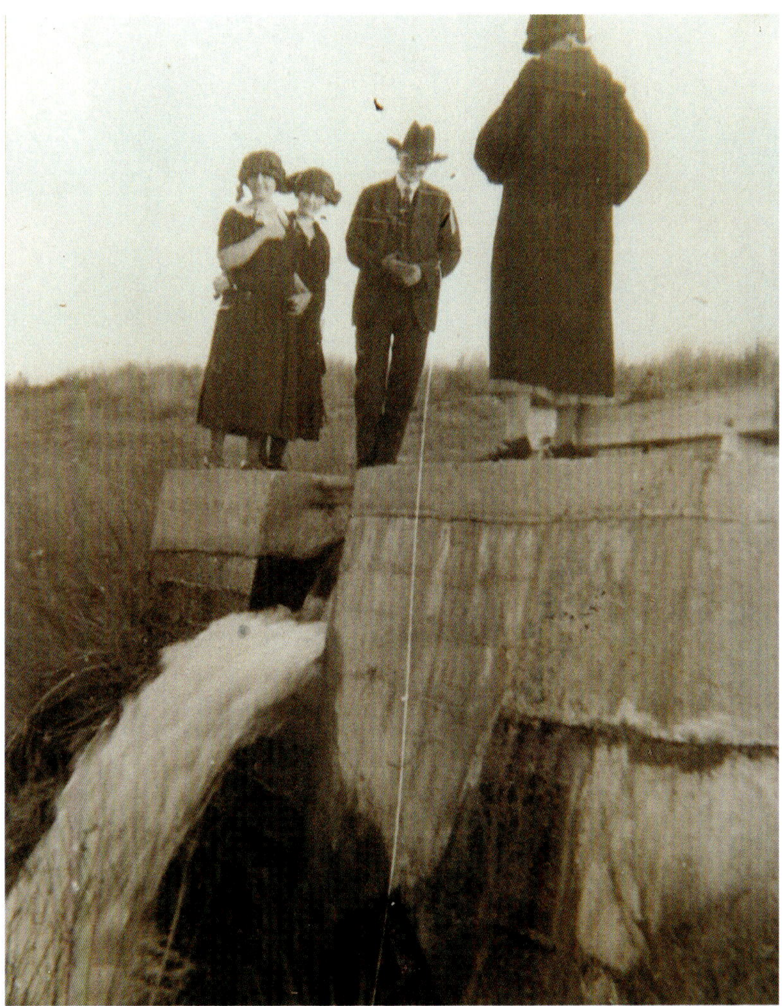

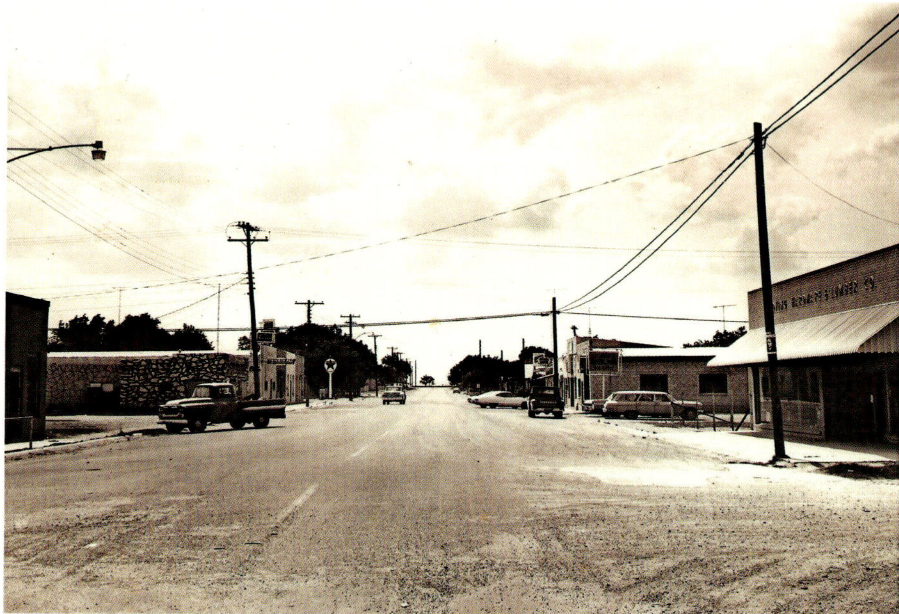

has also been petroleum production. Oil was discovered in 1921, which caused an increase in the population of the county. The modern population, which mostly lives in the county seat of Mentone, peaked at about six hundred in 1933 and has been declining ever since. A valuable asset for Loving County is Red Bluff Reservoir on the Pecos River just south of the Texas–New Mexico state line. This important facility not only impounds water for irrigating farms but also provides recreational activities such as fishing, sailing, and birding.

As U.S. 285 continues south from Loving County across Reeves County, the first identifiable town is Orla, now described as a ghost town. It has had a post office since 1906, but there are hardly any residents. A sign advertising the now-defunct Orla Grocery pretty well sums up the conditions of the area: "It's a long way to anywhere from Orla Grocery..." Arrows on the sign indicate the way to Pecos, El Paso, Jal, and Carlsbad.

Pecos is the county seat of Reeves County. The first Anglo settlers came into the area in 1871, but the first rush came in 1875 when word spread about the availability of open-range ranchland. The Texas and Pacific Railway was built from east to west across Reeves County in 1881, and Pecos sprang up around the depot. Reeves County was organized in 1884. Pecos is famous for holding the world's first rodeo on July 4, 1883.

Proceeding south from Carlsbad, U.S. 285 passes through the villages of Otis, Loving, and Malaga on the way to Loving County, Texas. These three little towns are in farming communities that have lost population as a result of the shrinking economic base.

Loving County has the dubious distinction of being the least populous county in the nation. Mostly it is ranching country, but there

The Pecos Valley Railway was constructed to Eddy in 1891, connecting with the Texas and Pacific at Pecos, and was extended to Roswell from Eddy in 1894. The railroad provided some economic stimulus and population growth, but the county's economy was dependent upon ranching and farming until petroleum production started in the 1920s.

The southernmost county that is covered in this book is Pecos County, where U.S. 285 passes through the county seat, Fort Stockton. This area was the locale of several historic trails—including the Comanche Trail—which crossed close to Horsehead Crossing and Comanche Springs. As early as 1840, travelers passed through the area on the Chihuahua Trail on the way from the Mexican state of Chihuahua to Santa Fe. Soon after the Civil War, the Goodnight-Loving Trail crossed present-day Pecos County. One of the most noted landmarks of the trail was Horsehead Crossing on the Pecos River.

The military outpost of Fort Stockton was established by the U.S. Army in 1858 at Comanche Springs. Its purpose was to protect the mail running between El Paso and San Antonio. The Butterfield Overland Mail started service to Fort Stockton the same year that the army post was established.

Fort Stockton—the town—started near the army post as Saint Gaul. The name was changed to Fort Stockton in the early 1880s. Pecos County was organized in 1875. There was some grain production in the county, but cattle and sheep ranching pretty much dominated the county's economy. In fact, La Escalera Ranch, which occupies large portions of Pecos, Reeves, and other counties, has

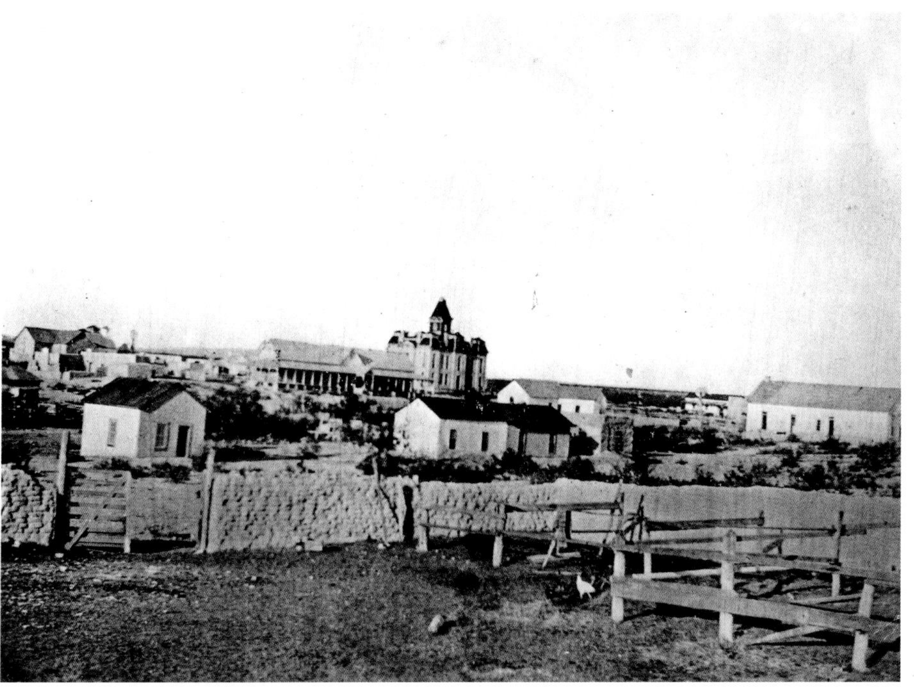

*Fort Stockton, Texas, undated. Courtesy Nita Stewart Haley Memorial Library, Midland.*

been in operation for about a century and, at 320,000 acres, is one of the largest ranches in the nation.

The population of Pecos County was stimulated in 1913 by the construction of the Kansas City, Mexico and Orient Railway across the

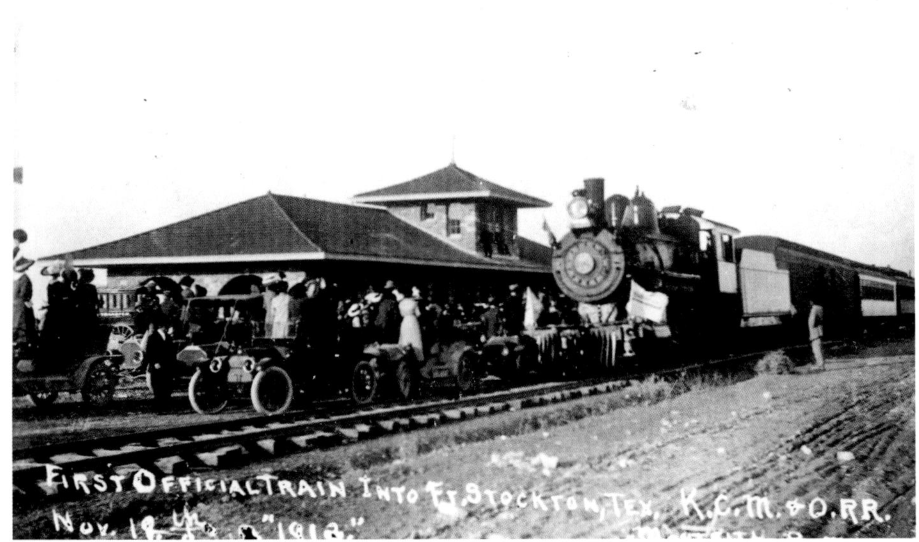


## Major Developments in the Mid-Twentieth Century

The entire U.S. 285 corridor was adversely affected by the Great Depression in the 1930s. Jobs disappeared as demand for farm crops and livestock plummeted. Businesses in the towns could not survive without the support of the agricultural economy. The principal exception to this was in the areas where the petroleum industry was just taking off before the depression hit. Oilmen continued to find reserves, build pipelines, and construct refineries. It was mostly independent operators who progressed with the smaller oil fields from 1926 to 1939; in 1939 Martin Yates II and his associates discovered the Loco Hills oil pool in Eddy County, the second largest in the nation. Despite the adverse effects of the depression, the economy of most of the area between Roswell and Fort Stockton benefited from the petroleum industry.

World War II was a tremendous blow to the people of the U.S. 285 corridor, as many of the young men left their jobs and families to enlist in the service. An important way in which the U.S. 285 corridor was affected by the war was the building of military bases in Roswell, Carlsbad, and Pecos. The Roswell Army Air Field opened in May 1942 as a flying school; bombardier training was soon added, and it became a full-fledged base. It continued after the war as home to the 509th Bomb Wing, the unit that had delivered the first atom bomb

county. The railroad caused a boom in land speculation as well as community growth, to which the irrigation projects along the Pecos River also contributed. The county's population more than doubled in the 1920s.

Petroleum production stabilized the economy in Pecos County, which is home to the Yates Oil Field, one of the largest oil fields in the United States. It was discovered in 1926, and jobs there led to the rise of the towns Iraan, Redbarn, and Bakersfield east of Fort Stockton.

in Japan and helped end the war. Roswell Army Air Field became Walker Air Force Base in January 1948, an important element of the Strategic Air Command. Then Walker became the Roswell Industrial Air Center in 1967.

Bases similar to the one at Roswell were also established in Carlsbad and Pecos. They operated from 1942 until the war ended in 1945. Besides training men to pilot war planes, the Pecos Army Air Field had a rather unique assignment: to train the Women Air force Service Pilots to fly military aircraft. The women flew several different models for engineering tests and for transporting freight. The Pecos Army Air Field was a sizable establishment, and its population rivaled that of the town during its peak of operations. After the war, the bases at Carlsbad and Pecos were devoted to civilian uses; for example, the Pecos Municipal Airport utilizes part of the former military base there.

World War II ended the depression and ushered in a period of limited prosperity that lasted several years. Business, industry, and agriculture all expanded and provided jobs for returning servicemen. It would be logical to speculate that many of the houses in this book were built during the period following World War II. A major part of that prosperity involved the growth of Walker Air Force Base and the associated growth of Roswell. The oil industry also grew during the postwar period in all the counties from Chaves to Pecos, attracting many new settlers seeking jobs. Another stimulus, especially in Reeves County, was the beginning of large-scale mining of sulfur in nearby Culberson County in the 1960s. This led to thirty years of sub-

stantial population and economic growth. The sulfur mining, plus the potash mined around Carlsbad, provided the principal cargo for the railroad in the Pecos Valley; passenger service was discontinued.

While the air base was in operation, Roswell was a thriving city. In the 1940s, its population almost doubled. The census of 1950 showed 40,605 residents in Chaves County, with 25,738 of those in Roswell. The 1960 count for the county was 57,649—the largest ever—with 39,593 in Roswell.

The first blow to the economy of Roswell was the deactivation of the missile squadron that had been operating within a twenty-five-mile radius of Walker Air Force Base. Roswell had already experienced the departure of several major petroleum companies. Then the worst event in the history of the U.S. 285 corridor as far as adversely affecting jobs and housing happened: the closure of Walker in 1967. Thousands of Roswell residents had to leave, and hundreds of them abandoned their houses. Whole neighborhoods became boarded-up ghost towns. Stores by the dozen closed their doors. Other towns in the U.S. 285 corridor suffered the negative impact of closing Walker, but none was as devastated as Roswell.

### The Twenty-First Century

North of Roswell, the U.S. 285 corridor tends to be a slowly changing world, where there are few towns and the ranch houses are far apart. However, the mechanization of agriculture and the urbanization of the United States have affected the area in modern times. For instance, ranchers ride the fences and ranges in pickup trucks and

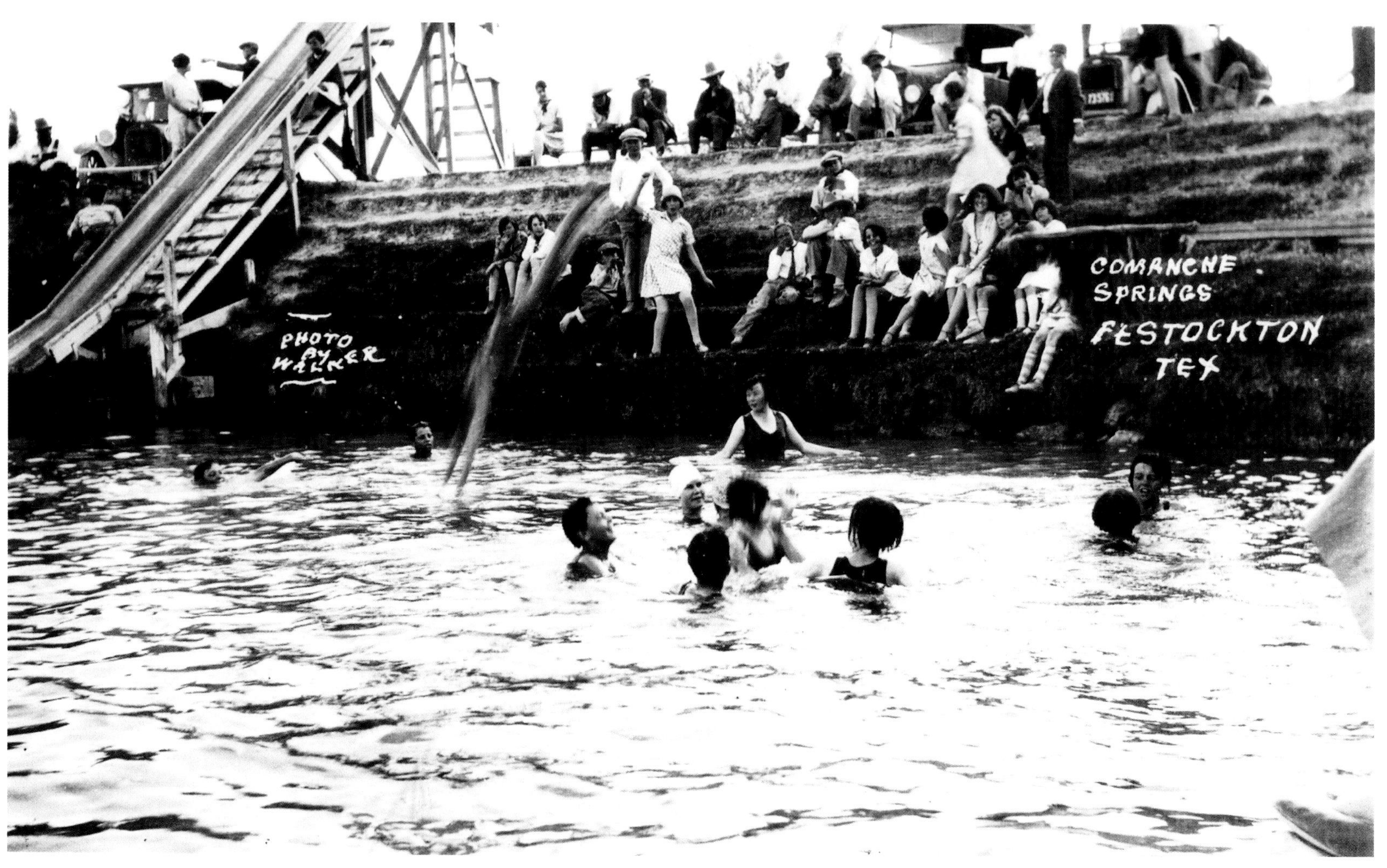


Comanche Springs, Fort Stockton, Texas. Courtesy Nita Stewart Haley Memorial
Library, Midland.

all-terrain vehicles, which are faster and more efficient than horses. Some ranchers do not even have horses. Other aspects of the work are mechanized as well, such as the mixing and distribution of feed to cattle, the chutes used for branding and doctoring livestock, and the big trucks that haul the stock to market. All of this contributes to a shrinking population in the ranching country; as fewer hands are needed on the ranches, more people head for the big cities to find suitable employment. James Cliett, who ranches north of Roswell, uses the volunteer labor of would-be cowboys from the city to help with the work. Cliett stated that he cannot afford to pay a living wage to cowboys.

Chaves County's main economic base is agriculture. A big change in the agriculture of the area started in the late 1980s when the number of dairies in the county suddenly grew to more than forty. The county ranks first among New Mexico counties and is the tenth-ranking county in the nation in milk production. Much of the local milk goes into mozzarella cheese at the Leprino Foods factory just outside of Roswell, said to be the largest such factory in the world. The county also leads New Mexico counties in the production of sheep, wool, and hay.

Closing the air base turned out to be a major turning point in the history of the Chaves County area and marked the beginning of an industrial period for the local economy. It started with the conversion of Walker Air Force Base into the Roswell Industrial Air Center (RIAC). The Roswell Industrial Development Corporation was created to attract business and industry to RIAC. The Roswell airport and numerous related businesses moved to RIAC, which had some of the longest runways in the world and provided an excellent testing facility for airlines and aircraft makers. The former military housing at the base was administered by the Roswell Housing Authority.

Over the next few years, RIAC saw the arrival of many manufacturing firms—making fireworks, Christmas ornaments, lollipops, jeans, mobile homes, and buses—some of which turned out to not be permanent, but they helped the city rebound.

A major development at RIAC came when Roswell Community College moved to the base and became Eastern New Mexico University–Roswell Campus. Over the years it outgrew the original buildings and constructed a beautiful new campus. The ever-growing college became a community university, offering educational advantages to local residents.

Roswell's recovery from the loss of population was slow but steady. A major factor in the recovery was an influx of retirees from the Northeast and Midwest, who responded to vigorous campaigns by the Roswell Retirement Service and other related organizations. Retirees were attracted not only by the low cost of living and the warm weather but also by the availability of nice houses at bargain prices.

The annual celebrations of the Roswell Incident—the 1947 alleged UFO crash—which started in 1995, have made Roswell a tourist destination. One important economic effect of the UFO Festival has been the establishment of several new motels by national chains. Roswell has been growing, gaining population and numerous new places of business; unfortunately mostly minimum-wage jobs are being created. The city has two world-class museums—the Roswell Museum and Art Center and the UFO Museum—and is the home of New Mexico's first state park, Bottomless Lakes. The annual Eastern New Mexico State Fair, serving thirteen counties, is the state's oldest fair and is second in size only to the New Mexico State Fair in Albuquerque. Main Street Roswell has been trying to revitalize downtown Roswell in the wake of numerous store closings after the opening of the Roswell Mall and the big-box stores. Some of the abandoned buildings in the Pecos Valley resulted from the establishment of malls and stores such as Wal-Mart and Target. While the big retailers provide jobs for many local people, they also tend to run the small merchants out of business.

Despite signs of growth and prosperity, a major challenge facing the people of the U.S. 285 corridor is unemployment and poverty. Roswell, the largest and oldest of the cities, has more than 20 percent of its population falling at or below the poverty level as defined by the federal government. The per capita income in Roswell is only about 45 percent of the national average, which means that more than a third of the children in Chaves County live at or below the poverty level. In most of Roswell's elementary schools, about 90 percent of the students qualify for free or reduced-price lunches. The unemployment rate was high for several years before the recession that started in 2007. It seems that most of the jobs created in Roswell since 1990 have been relatively low-paying positions that provide little stimulation to the local economy.

Artesia's main economic stimulus continues to be the petroleum industry. The Navajo refinery there is the largest in New Mexico. The petroleum production in this area is a mainstay of the budget of the state government. Tourists are attracted to the magnificent sculptures along Main Street depicting various aspects of the town's history and the petroleum industry and to the recreational facilities at Brantley Lake State Park south of town. An important employer in the community is the Federal Law Enforcement Training Center, which trains Border Patrol officers and others from all over the nation.

In Artesia, as in other towns along U.S. 285, the public schools provide jobs for local residents, and the Artesia schools are well known for their championship sports teams. Downtown has suffered the blight of most other small cities with businesses abandoning the area, but Main Street Artesia has accomplished some notable revitalization projects.

Carlsbad is a center for petroleum production, potash mining, and tourism. One of its major assets is a community college, which is a branch of New Mexico State University. Carlsbad Caverns National Park, located southwest of the town, is a world-class tourist attraction, and the two also profit from nearby Brantley Lake. In addition, Lake Carlsbad, near downtown on the Pecos River, draws visitors year-round. The lake provides water sports in the summer, and at Christmastime there are popular boat tours to view the superb light displays along the shore. The Living Desert Zoo and Gardens State Park is another gem offered to tourists. Like Roswell, Carlsbad owes much of its economic stability to the activities of retirees, who flock to the city because of the warm weather and many attractions.

Pecos, Texas, home of the first-ever rodeo, still puts on a rodeo every year in late June. Farming and ranching have declined, but the population is growing slowly and the public schools are reportedly doing well. Major employers in Pecos include the Reeves County government, the City of Pecos, schools, banks, a prison, and the local hospital. Despite many signs of life in Pecos, the extensive poverty in

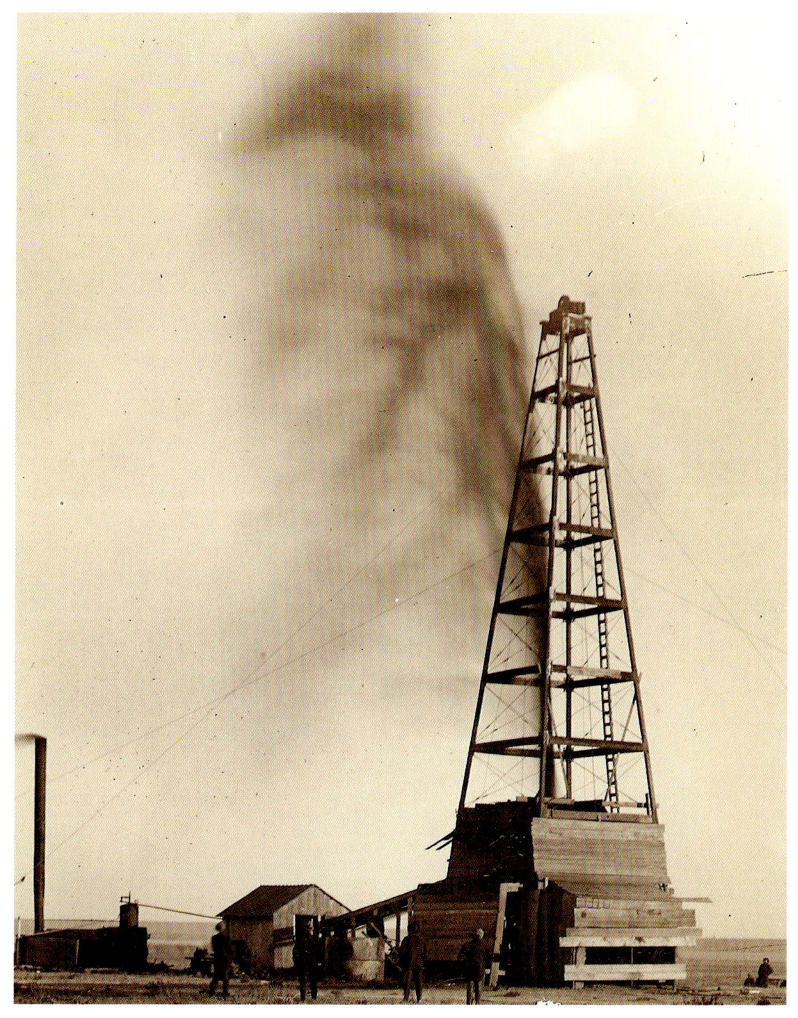

*Eddy County oil well, December 24, 1920. Courtesy HCSNM.*

the area makes it hard for businesses to thrive. Projects by the Main Street program to restore historic buildings in the downtown area have met with limited success, and most of the buildings are empty. The main claim to fame of Pecos is the cultivation of its famous cantaloupes, which are much desired by consumers.

Fort Stockton, Texas, is similar to Pecos in many respects but perhaps is in a bit better shape. Farms and ranches struggle sometimes but have been holding steady. The major employers in the area are oil and pipeline companies, public schools, banks, and two prisons. Interstate 10, not U.S. 285, is the major road through Fort Stockton. The town is considered the gateway to Big Bend National Park, and most visitors to the park pass through town on I-10. Another tourist attraction is the restored old Fort Stockton, and the Main Street program has been active in restoring the historic and downtown areas to their original appearance. But as in many other towns, it is very difficult to generate enough business for the historic areas to thrive. There is an updated visitor's center and a new community center. Fort Stockton's population is growing, albeit very slowly, and the surviving business owners feel somewhat content with their viability, but there is still widespread poverty.

### Conclusion

While the portion of the U.S. 285 corridor that lies north of Roswell is pretty much detached from the area south of Roswell to Fort Stockton, the southern area seems to be characterized by an attitude of success despite extensive poverty. Notwithstanding the problems brought on by difficult economic times, the five major cities in the area—Roswell, Artesia, Carlsbad, Pecos, and Fort Stockton—seem determined to fight for survival. These towns have seen a lot of ups and downs in the past century, as they have struggled to define who they are and what they want to accomplish for their residents.

The towns all had early growing pains in attracting initial settlers, roads, railroads, and businesses. They all weathered the Great Depression and World War II. They have constantly been on the alert when it comes to promoting developments that were beneficial to the residents and the state.

The smaller towns have survived the adverse economic conditions long enough to have some sense of security about the continued viability of their communities. At the same time, rampant unemployment and poverty have resulted in many broken dreams in the small towns.

So…the "houses where nobody lives" along the U.S. 285 corridor continue to keep watch over their closely guarded secrets. In the life of every one of the houses and stores in this study, a fateful day arrived when someone had to make a decision. The last occupant of each house had to decide that "today is the day that we must move away." The last businessperson to occupy each of the abandoned stores had to decide that "today is the day that we must close our doors for the last time."

Using the lens of the professional photographer, Chris Enos has endowed this book with the unmistakable touch of the artist. This delightful pictorial account of where the U.S. 285 corridor is situated in terms of the flow of history can help the residents of the area understand how their region fits into the scheme of things. Looking at the broken dreams that reflect where the area has been in the past helps the residents to assess not only where their communities are presently but also where they want to go.

**p. 2**
Vaughn, NM

HOUSES

**p.7**
Malaga, NM

**p. 8**
Ft. Stockton, TX    Ft. Stockton, TX
Encino, NM    Orla, TX

**p. 9**
Encino, NM    Pecos, TX
Pecos, TX    Pecos, TX

**p. 10**
Artesia, NM    Roswell, NM    Pecos, TX
Pecos, TX    Artesia, NM    Pecos, TX
Loving, NM    Vaughn, NM    Pecos, TX

**p. 11**
Ft. Stockton, TX    Pecos, TX    Pecos, TX
Pecos, TX    Encino, NM    Ft. Stockton, TX
Pecos, TX    Pecos, TX    Orla, TX

**p. 12**
Encino, NM

**p. 13**
Loving, NM
Pecos, TX

**p. 14**
Pecos, TX
Vaughn, NM

**p. 15**
Vaughn, NM

**p. 16**
Pecos, TX    Ft. Stockton, TX
Pecos, TX    Pecos, TX

**p. 17**
Artesia, NM    Pecos, TX
Artesia, NM    Ft. Stockton, TX

The publisher acknowledges these lenders to the publication: Historical Center for Southeast New Mexico (HCSNM); Palace of the Governors Photo Archives, New Mexico History Museum, Santa Fe; Nita Stewart Haley Memorial Library and History Center, Midland, Texas.

**Project editor:** Mary Wachs
**Art direction and production:** David Skolkin
**Design associate:** Jason Valdez
**Composition:** Set in Din
Manufactured in Singapore
10 9 8 7 6 5 4 3 2 1

Museum of New Mexico Press
Post Office Box 2087
Santa Fe, New Mexico 87504
www.mnmpress.org

Library of Congress cataloging-in-publication Data

Enos, Chris, 1944-

   Broken dreams : photographing southeast New Mexico to Texas / by Chris Enos ; essay by Elvis E. Fleming.

      p. cm.

   ISBN 978-0-89013-535-8 (clothbound : alk. paper)

1.  New Mexico--Pictorial works. 2.  New Mexico--History, Local--Pictorial works. 3.  Abandoned buildings--New Mexico--Pictorial works. 4.  Ghost towns--New Mexico--Pictorial works. 5.  Ranches--New Mexico--Pictorial works. 6.  Santa Fe Region (N.M.)--Pictorial works. 7.  Fort Stockton Region (Tex.)--Pictorial works.  I. Title.

   F797.E66 2011

   972--dc23

                          2011012538